AF564704

Fish Genetics and Biodiversity Conservation

Fish Genetics and Biodiversity Conservation

Abhiraj Jain

RANDOM PUBLICATIONS
NEW DELHI - 110 002 (INDIA)

Fish Genetics and Biodiversity Conservation

ISBN 978-93-51112-87-7

Published in 2014 in India by

RANDOM PUBLICATIONS

4376-A/4B, Gali Murari Lal, Ansari Road
New Delhi-110 002
Phone: +9111-43580356, 23289044
E-mail: randomexports@gmail.com; sales@randompublications.com; info@randompublications.com

Type Setting by: Friends Media, Delhi-110089
Printed at Thomson Press (India) Ltd

Preface

The genomics revolution and its impacts on aquaculture are expected to contribute to resolving problems such as diseases, environmental impacts, and low profit margins. The major potential applications of genome technologies, primarily in aquaculture but also to some extents in capture fisheries include: markerassisted selection (MAS) for genetic enhancement; environmetal improvements through enhanced productivity as well as the development of novel technologies for environment monitoring, development of effective vaccines and their delivery technologies; monitoring antibiotic resistance; diagnosis for fish diseases and for the safety of aquatic produce; accurate identification of fish stocks for capture fisheries management and for their use as FiGR in aquaculture; conservation of FiGR, including protection of endangered species, in response to fish production strategies and consumer interests; and the development and application of transgenic fish technology including, for example, sterilization technology to address concerns about their possible environmental impacts .

A great challenge for aquaculture and capture fisheries is the long-term conservation of FiGR. Genome technologies provide new tools for genetic analysis. Innovative DNA marker technologies have opened a broad avenue for the analysis of genetic diversity based on genotypes. Some aquaculture operations still use wild fish seed. For these and for future fish breeding programs, conservation of wild FiGR is important. The applications of genomics in aquaculture and capture fisheries raise ethical, economic, environmental, legal, and social concerns. The most prominent of these at present relate to the development and use of genetically modified organisms. More research is needed not only to resolve issues related to safety of using transgenic fish, but also to produce novel technologies allowing safe use of transgenic technology. Public education about genomics and its applications is a key issue. The public is relatively naïve and ill-

1

Fish Production: Structure and Composition

The fish production in India witnessed a spectacular growth since independence. It rose from a mere 0.75 million tonnes in 1950-51 to over 6.57 million tonnes in 2005-06. In the initial years, marine sector used to contribute more to total fish production then inland sector. In 1950-51, marine production contributed about 71.01%, which fell gradually to 42.77% in 2005-06, while inland sector started contributing from 28.99 in 1950-51 to about 57.23% in 2005-06. In fact, by the year 2000, its share crossed 50% and continues to improve its share further in the coming years.

Table: *Fish Production by Source in India*

Year	***Fish production (Mt)***		
	Marine	***Inland***	***Total***
1950-51	0.534	0.218	0.752
1960-61	0.880	0.280	1.160
1970-71	1.086	0.670	1.756
1980-81	1.555	0.887	2.442
1990-91	2.300	1.536	3.836
2000-01	2.811	2.845	5.656
2003-04	2.941	3.458	6.399
2004-05	2.779	3.526	6.305
2005-06	2.816	3.756	6.572

Expansion of fleet capacity, technological innovation, and increases in investment all led to explosive growth in the exploitation of marine fisheries through the 1960s, 1970s and 1980s. But from the late 90s onwards, the marine fisheries production reached a plateau and it seems that it can register only marginal increase in the near future. With most wild fisheries near maximum sustainable exploitation levels, capture fisheries will most likely grow slowly.

On the other hand, inland fish production was on constant rise. The inland fisheries include both capture and culture fisheries. The capture fisheries have been the major sources of inland fish production till mid-1980s.

But the fish production from natural waters like rivers, lakes, etc. followed a declining trend, primarily, due to proliferation of water control structures such as dams and habitat degradation due to human interference and settlement near and surrounding water bodies (Katiha and Bhatta, 2002).

The depleting resources, energy crisis and resultant high cost of fishing etc. have led to increased realisation of the potential and versatility of aquaculture as a sustainable and cost-effective alternative to capture fisheries.

Trends and Current Patterns of Fish Production

To assess the trend of fish production in the country, the last 25 years data (from 1980-81 to 2005-06) was used. Both linear and non-linear trend analysis was tried and the best equation fitted. The total fish production in India has been following a linear trend and it is likely proceed further in the same direction.

The trend pattern of marine and inland fish production revealed that while the marine sector's production was increasing at a decreasing rate, the inland sector's production was increasing at an increasing rate, the possible reasons for which are that inland aquaculture activities are gaining much importance in some of the states like Uttar Pradesh, Andhra Pradesh, Punjab etc. in recent years. The trend equations for both marine and inland fish production, as well as for total fish production are given below:

Marine: $Y = -2.7588X + 143.4X + 1061.3$

Inland: $Y = 2.9078X + 38.083X + 824.85$

Total: $Y = 185.5X + 1867.4$

Based on the trend observed in the last 25 years, assuming that the same scenario would prevail in both public and private investment pattern in the sector, the fish production of the country would reach 7.46, 8.42 and 9.38 million tonnes in 2010, 2015 and 2020 respectively.

***Table:** Fish Production in 2020*

Year	(in million tones)		
	Marine	Inland	Total
2010	2.88 (38.61)	4.58 (61.37)	7.46
2015	2.70 (32.07)	5.72 (67.33)	8.42
2020	2.38 (25.37)	7.00 (74.63)	9.38

Note: Parentheses indicates percentage to total

In the coming decades, aquaculture will likely be the greatest source of increased fish production, as fish farmers expand the water surface area under cultivation and increase yields per unit of area cultivated. But the sector must overcome several major challenges, if it has to sustain the rapid growth of the past 25 years:

- Aquaculture has to face competition from others users of land and water, as these resources would become more scarce in future (Rosegrant *et al*, 2002)
- Marine sector would face energy crisis, as it would need more fossil fuels for further expansion
- The aquaculture production would be restrained by diseases as the sector expands (Subasinghe, *et al*, 2001)
- The availability of fish meal, fish oil and wild caught trash fish as feed inputs may also become a limiting factor.

Yield increases can come either from increased inputs or greater efficiency of inputs. It is likely that in the next several decades, aquaculture production will benefit from both these sources of yield growth. Greater use of compounded aquaculture feeds along with improvements in rearing technology and selective breeding has the potential to significantly increase the productivity of many forms of aquaculture. In the last 25 years, total fish production has been growing at an annual growth rate of about 4.60%, in which marine sector was growing at a rate of 3.24 and inland sector was growing at a rate of 6.20%.

***Table:** Compound Growth Rate in Fish Production, 1980-81 to 2005-06*

Year	Marine	Inland	Total
1980-81 to 1989-90	3.80	5.28	4.39
1990-91 to 1999-00	2.33	6.55	4.13
2000-01 to 2005-06	-0.21	5.37	2.75
1980-81 to 2005-06	3.24	6.20	4.60
1980-81 to 1980-91 (Pre-WTO)	4.35	5.43	4.78
1991-92 to 2005-06 (Post-WTO)	0.84	5.71	3.18

Overall, inland sector fared better in all the periods, viz. 1980-81 to 1989-90, 1990-91 to 1999-00 and 2000-01 to 2005-06. There seems to be a slower pace in growth of this sector in the recent times. In contrast, marine sector witnessed a negative growth rate in the period 2000-01 and 2005-06, which indicates the exhaustion of marine resources especially in the in-shore and near-shore waters, where maximum harvesting has been carried out. About 90% of the present production from the marine sector is within a depth range of up to 50-70 m and the remaining 10% from depths extending upto 200m. While 93% of the production is contributed by artisanals and motorized sectors, the remaining 7% is contributed by deep sea fishing fleets confining their operation mainly to the shrimp grounds in the upper east coast. Hence, in order to enhance and sustain the contribution from this sector, we need to go into the deep sea for targeting its untapped potential. This requires enhanced investment in mechanized vessels, capacity strengthening of artisanals and probably a proper institutional structure to share the benefits.

The growth rates in pre and post-WTO periods were also estimated. It is noticed that the pre-WTO period witnessed an impressive growth rate of about 4.78 as compared to post WTO period (3.18). This trend was mainly due to the marine sector, which is understandable by the fact that the country's fish export basket was dominated by marine species and buoyance of marine export might have propelled the growth of marine catch, and vice versa. The post-WTO period imposed many quality regulations in terms of SPS measures on developing countries like India, which couldn't create huge investment in the infrastructures required to produce export-quality marine fisheries products that are acceptable to our trading partners, especially EU, USA and Japan. In contrast to the marine sector, inland sector continued to grow better in the post-WTO period also, which is possibly because of enhanced public and private investment in inland fisheries sector especially though different development programmes and research by the Government of India since IV plan, that started delivering results continuously.

Figure clearly depicts the dominance of inland fisheries sector in the contribution to total fish of the country, particularly from 1990s.

Drivers for Future Growth

To sustain this growth of the sector in general; technology, infrastructure and market would play a major role apart from enhanced investment in research and development. Technology had been the main factor responsible for the phenomenal growth of aquaculture, particularly after the advent of carp polyculture and composite fish culture in the late 70s.

Similarly, major investments on infrastructure such as construction of mini harbors, jetties, landing centres, introduction of trawlers and mechanized vessels, supply of nets, etc. led to increased catch and contribution from capture fisheries sector. However, market has not been able to play the role of a major driver for the growth of the sector so far. To un-tap the potential of the sector, market should take the lead in furthering the growth, especially in the emergence of aquaculture sector. Some of the illustrative aspects under each major driving force are given below for future attention:

1. Technology
    1. Quality seed production
    2. Selective breeding of carps
    3. Formulation of low cost feed materials using locally available ingredients
    4. Fabrication of nets for targeted fishing
2. Infrastructure
    1. Construction of landing centres in second tier potential costal towns
    2. Creation of cold storage near landing centres
    3. Up-gradation of manually operated boats into outboard motorized ones
    4. Supply of ice box

Opportunities in Marine Biotechnology and Aquaculture

The oceans offer abundant resources for research and development, yet the potential of this domain as the basis for new biotechnologies remains largely unexplored. Indeed, the vast majority of marine organisms (primarily microorganisms) have yet to be identified. Even for known organisms, there is insufficient knowledge to permit their intelligent management and application. Oceanic organisms are of enormous scientific interest, for two major reasons. First, they constitute a major share of the Earth's biological resources. Second, marine organisms often possess unique structures, metabolic pathways, reproductive systems, and sensory and defence mechanisms because they have adapted to extreme environments ranging from the cold polar seas at -2° C to the great pressures of the ocean floor, where hydrothermal fluids spew forth. Most major classes of the Earth's organisms are primarily or exclusively marine, so the oceans represent a source of unique genetic information.

The oceans therefore offer abundant resources for research and development. Yet the potential of this domain as the basis for new biotechnologies remains largely unexplored. Indeed, the vast majority of marine organisms (primarily microorganisms) have yet to be identified. Even for known organisms, there is insufficient knowledge to permit their intelligent management and application.

Interest in marine biotechnology has been growing in recent years, (2) By contrast, the U.S. Government invested less than $44 million on marine biotechnology R&D in fiscal year 1992, (4). Only about 1 or 2 percent of the total Federal investment in biotechnology has been devoted to marine biotechnology and aquaculture. Even so, promising results are emerging from these activities. Additional Federal support of research in key areas of marine biotechnology and aquaculture will generate both new fundamental knowledge and advanced technologies for producing new pharmaceuticals, biomaterials, and other products; developing and improving bioremediation and bioprocessing; enhancing cultivation of aquatic species; and expanding understanding of biological processes in the oceans and their role in global change.

This chapter provides an overview of recent biotechnology applications in marine science and aquaculture and identifies research priorities and specific opportunities that offer potential for significant advances. The chapter emphasizes development of salt-water resources but includes biotechnology applications involving fresh water, including thermal springs.

The five research priorities are:

- Develop a fundamental understanding of the genetic, nutritional, and environmental factors that control the production of primary and secondary metabolites in marine organisms, as a basis for developing new and improved products.
- Identify bioactive compounds and determine their mechanisms of action and natural function, to provide models for new lines of selectively active materials for application in medicine and the chemical industry.
- Develop bioremediation strategies for application in the world's coastal oceans, where multiple uses — including wastewater disposal, recreation, fishing, and aquaculture — demand

prevention and remediation of pollution; and develop bioprocessing strategies for improving sustainable industrial processes.

- Use the tools of modern biotechnology to improve the health, reproduction, development, growth, and overall well- being of cultivated aquatic organisms; and promote the interdisciplinary development of environmentally sensitive, sustainable systems that will enable significant commercialization of aquaculture.
- Improve understanding of microbial physiology, genetics, biochemistry, and ecology in order to provide model systems for research and production systems for commerce, and to contribute to understanding and conservation of the seas.

New and Improved Products from The Seas

PRIORITY: Develop a fundamental understanding of the genetic, nutritional, and environmental factors that control the production of primary and secondary metabolites in marine organisms, as a basis for developing new and improved products.

PRIORITY: Identify bioactive compounds and determine their mechanisms of action and natural function, to provide models for new lines of selectively active materials for application in medicine and the chemical industry. Most major groups of living organisms are primarily or exclusively marine. Tropical marine environments harbor an especially wide diversity of animals and plants. Many marine organisms are sessile and must employ sophisticated methods to compete for a place to anchor. This characteristic is reflected in part by a metabolism that produces enormously diverse bioactive products — many with no terrestrial counterparts.

Recent research has uncovered unicellular and multicellular microorganisms that are unique to the marine world. Indeed, "... it seems clear that marine bacteria are emerging as a significant chemical resource" and the amounts and characteristics of chemicals they produce.

Federal support for research in this area is essential, both to answer the fundamental questions and to assure that the resulting knowledge is translated into sustainable technologies. While it will be vital to cultivate marine microorganisms that produce novel products, the alternative approach of transferring genes of interest into non-marine microorganisms also should be investigated. For example, the capability to produce a marine polysaccharide — a complex molecule that could be useful as a food additive or a water-resistant adhesive — could be transferred to an easily grown bacterium (e.g., *E. coli* or *Bacillus subtilis*). This approach might be more effective in some cases than would cultivating the marine organism or recreating artificially the long and complicated production pathway for the polysaccharide.

Pharmaceuticals

Many bioactive substances from the marine environment already have been isolated and characterized, several with great promise for the treatment of human diseases. The compound manoalide from a Pacific sponge, for example, has spawned more than 300 chemical analogs, with a significant number of these going on to clinical trials as anti-inflammatory agents.

To promote sustainable technology development, the Federal Government should support research leading to new methods for discovery. One approach would be to focus on learning about natural functions, regulation, and production of substances generated by marine organisms, in order to identify potentially important agents. Refined test systems should be developed in order to identify selective agents produced by marine organisms.

Rapidly developing assay technology can facilitate exploration of the bioactivity of newly discovered compounds. These assay methods, which employ specific receptors for known physiological agents, require only minute amounts of a test substance and can be automated. (Traditional chemical tests require considerable amounts of test material and labourious measurement processes.) The new tools will make it feasible to test newly discovered compounds rapidly by the hundreds, for a wide spectrum of biological activities.

To date, exploitation of natural agents from the sea has been hindered by problems with limited or sporadic distribution and production. Much more research must be conducted to determine what seasonal factors and life cycle or reproduction states are linked with natural production of an agent. Factors influencing production may include diet, physical and chemical conditions, distribution by phylogenetic affiliation, geographic location, water depth, or associations with symbiotic microorganisms. Knowledge of these factors will be important in developing methods for producing selected metabolites, either from whole organisms or *in vitro* from cell or tissue cultures of plants and animals. Many of these compounds are very large and complex molecules, requiring very elaborate biochemical processes; as a result, it can be difficult to synthesize them or clone all the genes for standard production through fermentation.

Enzymes

Enzymes produced by marine bacteria are important in biotechnology due to their range of unusual properties. Some are salt-resistant, a characteristic that is often advantageous in industrial processes. The extracellular proteases are of particular importance and can be used in detergents and industrial cleaning applications, such as in cleaning reverse-osmosis membranes. *Vibrio* species have been found to produce a variety of extracellular proteases. *Vibrio alginolyticus* produces six proteases, including an unusual detergent-resistant, alkaline serine exoprotease. This marine bacterium also produces collagenase, an enzyme with a variety of industrial and commercial applications, including the dispersion of cells in tissue culture studies.

Other research has demonstrated the presence in algae of unique haloperoxidases (enzymes catalyzing the incorporation of halogen into metabolites). These enzymes could become valuable products, because halogenation is an important process in the chemical industry. Japanese researchers have developed methods to induce a marine alga to produce large amounts of the enzyme superoxide dismutase, which is used in enormous quantities for a range of medical, cosmetic, and food applications.

An unusual group of marine microorganisms from which enzymes have been isolated are the hyperthermophilic archaea (previously called archaebacteria), (9) which can grow at temperatures over 100° C and therefore require enzyme systems that are stable at high temperatures. Archaea typically are found in extreme environments, such as hot springs, animal guts, hydrothermal vents, sewage sludge digesters, and hypersaline habitats, including the Great Salt Lake.

Thermostable enzymes offer distinct advantages, many still to be discovered, in research and industrial processes. Thermostable DNA-modifying enzymes, such as polymerases, ligases, and restriction endonucleases, already have important research and industrial applications. Hot springs in Yellowstone National Park provided the first archaeon (*Thermus aquaticus*) from which thermostable DNA polymerases were isolated. These novel enzymes (the *Taq*® polymerases) became the basis for the polymerase chain reaction (PCR), a useful technique for studying genetic material. In 1989, thermostable DNA polymerase was designated Molecule of the Year by Science magazine. Comparable enzymes continue to be discovered. Most enzymes involved in the primary metabolic pathways of thermophilic bacteria and archaea are dramatically more thermostable than are their counterparts living at moderate temperatures. Expanded study of enzymes from thermophilic marine microorganisms will contribute to the understanding of mechanisms of enzyme thermostability and should enable the identification of enzymes suitable for industrial applications as well as modification of enzymes to enhance thermostability.

Biomolecular Materials

Recent research has demonstrated that marine biochemical processes can be exploited to produce new biomaterials. For example, a corporation in Chicago is commercializing a new class of biodegradable polymers modelled on natural substances that form the organic matrices of mollusk shells. Equally exciting are the mechanisms used by marine diatoms, coccolithophorids, mollusks, and other marine invertebrates to generate elaborate mineralized structures on a nanometre scale (less than a billionth of a metre in size).

Nanometre-scale structures can have unusual and useful properties. Research that will enhance understanding and allow engineering of the processes for creating these bioceramics promises to revolutionize the manufacturing of medical implants, automotive parts, electronic devices, protective coatings, and other novel products. Biomaterials also hold promise for counteracting biofouling, which long has been recognised as an extensive and costly problem. Bacterial biofilms form slime layers that increase drag on moving ships, interfere with transfer on heat exchangers, block pipelines, and contribute to corrosion on metal surfaces. Bacterial and microalgal colonization of surfaces is accompanied by settlement of invertebrate larvae and algal spores, eventually leading to "hard fouling" and the need for costly cleaning.

The most effective anti-fouling coatings have utilised toxic chemicals, such as copper and organotins. There is an urgent need for non-toxic biofouling control strategies, due to heightened recognition of the impact that toxic coatings can have on the environment. Research is needed on the attachment mechanisms of marine organisms and the natural products they employ to prevent fouling of their own surfaces. Molecular approaches to characterizing biofilm structure and development offer considerable potential for finding novel biofouling prevention strategies. It is now possible to determine the genes and pathways involved in regulation and synthesis of bacterial adhesive polymers. Considerable progress has been made in understanding the nature and expression of surface polymers produced by microorganisms such as the nitrogen-fixing *Rhizobium* species and the opportunistic pathogen *Pseudomonas aeruginosa*. Similar approaches can been applied to marine biofilm bacteria, to find the genetic determinants of adhesive production and the environmental factors that regulate synthesis.

Molecular biology techniques also can be applied to determine the basis of natural antifouling mechanisms. Many marine plants and animals remain free of attached bacteria, either because they produce repelling compounds or because their surface structure neutralizes bacterial adhesives. A product generated by the seagrass *Zostra marina* (eelgrass), for example, is an effective agent for preventing fouling by bacteria, algal spores, and a variety of hard-fouling barnacles and tube worms. Molecular characterization of natural fouling resistance could provide new strategies for fouling control. Potential applications include prevention of fouling in industrial pipelines or heat exchangers, improved design of trickling filters or aquaculture circulation systems, and control of biofilm infections of medical implants and prosthetic devices.

Biomonitors

Marine organisms can provide the basis for development of biosensors, bio-indicators, and diagnostic devices for medicine, aquaculture, and environmental monitoring. One type of biosensor employs the enzymes responsible for bioluminescence. The *lux* genes, which encode these enzymes, have been cloned from marine bacteria such as *Vibrio fischeri* and transferred successfully to a variety of plants and other bacteria.

The *lux* genes typically are inserted into a gene sequence, or operon, that is functional only when stimulated by a defined environmental feature. The enzymes responsible for toluene degradation, for example, are synthesized only in the presence of toluene. When *lux* genes are inserted into a toluene operon, the engineered bacterium glows yellow-green in the presence of toluene. This genetically engineered system "reports" that biodegradation of a specific chemical, in this case toluene, is proceeding. Another type of biomonitor that holds great promise is the gene probe, which can be used to identify organisms that pose health hazards or may be useful in research. Specific gene probes can be employed, for example, to detect human pathogens in seafood and recreational waters; fish pathogens in aquaculture systems; microorganisms capable of mediating desired chemical transformations (e.g., toxic chemical degradation, CO2 assimilation, metal reduction); and specific fish stocks in fish migration and recruitment studies.

Biopesticides

Natural marine products have the potential to replace chemical pesticides and other agents used to maximize crop yields and growth. Continued Federal support for R&D in this area is likely to result in useful natural pesticides that would provide greater specificity and fewer harmful side effects than do conventional synthetic agents. Current U.S. expenditures for all pesticides amount to $47 billion annually; by the year 2000, biopesticides from marine and other sources are expected to capture an estimated 10 percent of this market.

An example of a marine biopesticide in use today is Padan*TM*, which was developed from a bait worm's toxin known to ancient Japanese fishermen. This natural pesticide has demonstrated activity against larvae of the rice stem borer, the rice plant skipper, and the citrus leaf miner, among other pests. More recently, scientists in Montana discovered novel compounds in marine algae and marine sponges containing symbiotic microorganisms. These compounds promoted growth and stimulated germination and increased root and coleoptile lengths in test plants.

Several sponge and nudibranch species produce terpenes, a broad class of aromatic compounds used in solvents and perfumes and known to deter feeding by fish. Extracts derived from these same sponge and nudibranch species also demonstrated powerful insecticidal activity against two species, grasshoppers and the tobacco hornworm.(13)

Biomass for Energy Production

Approximately 40 percent of all primary energy production, or photosynthesis, occurs in the seas. In this process, oceanic plants (phytoplankton, seaweeds, seagrasses) take up carbon dioxide (CO*2*) and, with light energy from the sun, convert it into organic carbon (primarily sugars) and oxygen. The oceans contain 50 times as much carbon dioxide as does the atmosphere, and it is estimated that primary production incorporates 35 gigatons (1 gigaton = 1 x 1015 grams) of carbon into marine biomass annually. This abundant source of fuel for energy production has not been tapped commercially because it is not competitive with soybean meal and other easily harvested, traditional sources of biomass, and also because, regardless of the source, biomass is not competitive with other types of fuels.

The Federal Government should continue to support research on the use of biotechnology to enhance biomass production and utility. At least three general approaches are being explored.

First, the enzyme that captures CO_2 for photosynthesis — ribulose bisphosphate carboxylase\oxygenase or "RUBISCO" — is relatively inefficient, so supercomputers are being used to verify structural information, and the enzyme is being redesigned to optimize its function. Second, the chemical composition of biomass can be altered to make it more suitable for particular applications. For example, marine microalgae are being genetically engineered to boost their lipid content, with the aim of providing a source of alternative fuels that is more economical than are conventional sources. Third, biotechnology is being used to convert biomass to ethanol and other alternative forms of energy and chemical feedstocks.

Aquaculture

PRIORITY: Use the tools of modern biotechnology to improve the health, reproduction, development, growth, and overall well-being of cultivated aquatic organisms; and promote the interdisciplinary development of environmentally sensitive, sustainable systems that will enable significant commerciali-zation of aquaculture.

Aquaculture, which long has been practiced in Asia and is increasingly popular in the United States, Europe, and South America, will benefit tremendously from the use of new molecular tools and processes. With worldwide seafood demand projected to increase 70 percent in the next 35 years, and harvests from capture fisheries stable or declining, aquaculture will have to produce seven times as much seafood as it generates now to supply global demand by 2025. The use of modern biotechnology to intervene in the rearing process and enhance production of aquatic species holds great potential not only to meet this demand, but also to improve U.S. competitiveness in aquaculture.

The U.S. aquaculture industry has grown rapidly in recent years. Farm gate receipts exceeded $800 million in 1992 — a fourfold increase since the early 1980s.

The growth and international competitiveness of the U.S. aquaculture industry will be determined by the size of the resource investment in research and technology development. This investment should be made through a partnership of Federal and state agencies and the private sector. The Federal role is to provide leadership in supporting research to advance knowledge in important research areas and to facilitate the transfer of promising results and technologies to the private sector.

The major research issues in aquaculture are similar to those for other agricultural sectors, but the knowledge base for aquaculture is comparatively meager. Development of this knowledge is a particular challenge due to the diversity of cultured aquatic species and the systems for their production. Federal support for biotechnology research in this area will expand the knowledge base and yield significant dividends. The application of biotechnology promises significant benefits to both producers and consumers of aquacultural products. The use of genetically enhanced organisms may improve production efficiency through improvements in growth rates, food conversion, disease resistance, and product quality and composition. The application of biotechnology to aquaculture also may help conserve wild species and genetic resources and provide unique models for biomedical research.

Enhancing Reproduction and Early Development

Biotechnology can be applied to enhance reproduction and early development of cultivated aquatic organisms. The resulting benefits could include year-round production of gametes and fry of economically valuable species and creation of new markets for specialized, genetically improved broodstock. Similarly, biotechnology may provide techniques for improving the reproductive success and survival of endangered species, thereby helping to preserve the diversity of life on Earth.

As a first step, research should be directed toward improving basic understanding of environmental, hormonal, biochemical, and genetic control of reproduction. More specifically, scientists must identify and understand the mechanisms of expression of genes involved in reproduction and development, improve technologies for cryo-preserving gametes and embryos, improve delivery systems for administration of natural and synthetic hormones, and enhance understanding of the pharmacokinetics of uptake and release of administered hormones.

Improving Health and Well-Being

Biotechnology offers substantial opportunities to improve the health and well-being of cultivated aquatic organisms. More than 50 diseases affect fish and shellfish cultured in the United States, causing losses of tens of millions of dollars annually. Biotechnology not only can improve the survival, growth, vigor, and well-being of cultivated stocks, but also can reduce disease transfer between cultivated and wild stocks. New products and market opportunities can be developed related to aquatic animal health and well-being.

The tools of molecular biology can provide a basic understanding of host immunity, resistance, and susceptibility to diseases and associated pathogens by furnishing information about life cycles and mechanisms of pathogenesis, antibiotic resistance, and disease transmission. Improved technologies must be developed for detecting and diagnosing pathogens and diseases and for enhancing the genetic basis of disease resistance, thereby reducing the need for antibiotics and other drugs. Potential products resulting from this research include gene therapy techniques; broodstock free of pathogens; safe, effective prophylactic agents, including immune modulators, antigens, and vaccines; safe, effective therapeutic agents; and improved systems for administering prophylactic and therapeutic agents.

Improving Quality and Value

The Federal Government has a responsibility to help ensure the safety and quality of food supplies, and biotechnology can and should be an invaluable tool in carrying out this mandate. Biotechnology can be employed to assess and improve the safety, freshness, colour, flavour, texture, taste, nutritional characteristics, and shelf life of aquacultural food products. In addition, practical technologies can be developed to detect and assay toxins, contaminants, and residues in seafood, and to reduce or eliminate contaminants. There are also opportunities to apply biotechnology in improving seafood processing. Research should be conducted to develop and improve technologies for all these applications.

Conserving Genetic Resources

The preservation and enhancement of biodiversity in natural systems is an important Federal priority. Therefore, the Federal Government should encourage and support programs to maintain and enhance biodiversity in aquatic systems through cultivation and stocking of aquatic species. Biotechnology can be employed in two ways to conserve genetic resources of aquatic species.

First, the tools of biotechnology can be used to identify and characterize important aquatic germplasm, including endangered species. Genomes of aquatic species can be analysed and characterized, and quantitative trait loci identified. Second, biotechnology can be applied to improve understanding of the molecular basis of gene regulation and expression as well as sex determination and thereby improve methods for defining species, stocks, and populations. Approaches include developing marker- assisted selection technologies, improving precision and efficiency of transgenic techniques, and improving technologies for the cryo-preservation of gametes and embryos. Ultimately, stocking certain areas with selected, cultivated species and strains could help maintain biodiversity in natural aquatic ecosystems.

Enhancing Biomedical Models

Aquaculture has important purposes other than food production. Because they often adapt to extreme environments, marine organisms can provide unique models for research on biological and physiological processes. Studies of the developmental, cellular, and molecular aspects of marine organisms as model systems will provide insights into the basis of disease mechanisms and pathogenesis in humans. By contrast, use of mammalian organisms as a basis for the development of some types of human disease models may be neither feasible nor cost effective.

Progress in this arena will require that sophisticated molecular biology technologies be adapted to marine organisms, in order to enhance understanding of their biological processes. For example, approaches for gene transfer into eggs have been developed for many terrestrial organisms, but not for most marine species. This technology is needed for analyses of gene regulatory systems and gene expression. In addition, methods need to be developed for culturing tissues from marine organisms. Cultured cell lines will provide opportunities for gene transfer and gene expression studies and enhance the usefulness of marine species as biomedical research models. This is an important research area deserving of Federal support.

Understanding and Conserving the Seas

PRIORITY: Improve understanding of microbial physiology, genetics, biochemistry, and ecology in order to provide model systems for research and production systems for commerce, and to contribute to understanding and conservation of the seas.

Scientists have a powerful new array of sampling devices and measuring instruments that will accelerate greatly the acquisition of knowledge about ocean resources and foster their wise use. These technologies include manned deep-sea subme-rsibles, remotely operated vehicles, geosynchronous satellites, sophisticated acoustic measuring devices, pressure- retaining deep-sea samplers, geographic information systems, real-time flow cytometry, PCR and biomonitoring techniques, compu-terized databases, and other forms of information exchange and analysis. These tools should be exploited to accelerate the discovery of unknown marine microorganisms and to expand understanding of known varieties. Federal support for this research is essential, because only then will sufficient information be acquired to assure that practical applications will result. As new life forms and processes become known, and as understanding of them grows, marine biotechnology will make significant contributions to the nation's social and economic well-being.

Identifying Organisms and Their Niches

Biodiversity

Nowhere in the biosphere is biological diversity greater than in the seas, and the extent of this diversity becomes increasingly evident as scientists investigate new environments. In the 1980s, for example, giant tube worms (*Riftia*) were recovered from areas adjacent to deep-ocean thermal vents, and novel mussels (*Bathymodiolus*) that farm methanotrophic bacteria on their gill tissue were discovered around methane seeps in the Gulf of Mexico. Most newly described species have been and will continue to be microorganisms, although it is clear that new marine plants and animals also await discovery.

Less than 1 percent of the extant bacteria — marine and terrestrial — have been isolated and described.(20) The marine environment represents a particularly fertile source for new bacteria, as evidenced by the recent discovery of unusual "cold water" archaea 100-500 metres deep in the oceans. These archaea comprise a high percentage of the total bacterial ribosomal RNA present in seawater samples, yet they have not been isolated in pure culture and described. Pursuit of this research may provide new understanding of oceanic processes and, once these archaea are cultivated, perhaps a novel source of products. Countless other marine bacteria also have yet to be cultured; advances in microbial culture technology will enable scientists to isolate, describe, and possibly exploit these organisms.

Viruses are another newly appreciated element of marine biological diversity. Electron microscopy and PCR technology have revealed abundant numbers of viral particles in seawater samples. Specific viruses that infect species of marine phytoplankton have been cultured from coastal as well as open ocean sites. This discovery is exciting for at least two reasons. Viral infection of higher forms of marine life almost certainly affects global ocean processes, such as photosynthesis. Marine viruses also will provide new materials for development of genetic and biotechnological tools that can be used to study and manipulate marine organisms. For example, marine viruses could be used to genetically engineer higher forms of marine life.

Marine Ecology

As a basis for developing new applications for marine products and processes, analysing global climate change, and improving fisheries management, scientists must build a fundamental understanding of marine organisms and their specific adaptations to and interactions with their environment. Recent developments in molecular techniques are rapidly expanding capabilities for research on marine ecology. Three research areas seem to be especially fertile and deserving of Federal support.

As a first step, the new molecular tools should be applied to gain insight into the basic molecular and cellular processes by which marine organisms adapt to extreme environments. Examples of basic research that may lead to commercial applications are gene sequencing projects and the ongoing study of special glycoproteins that inhibit ice crystal growth in the tissues of Antarctic fish. These substances, along with gene sequence information for key marine species, may prove useful in industrial and medical preservation processes.

Second, a thorough understanding of marine ecological systems must be developed in order to specify the "normal" baseline level of function and to monitor and predict potential changes and perturbations of systems due to physical, chemical, or biological impacts. The development of predictive models for analysing potential global climate changes depends on the acquisition of fundamental information on molecular regulatory mechanisms of photosynthesis in the oceans.

Third, research on marine ecology can be conducted to benefit fisheries management. The tools of biotechnology can be used to determine the effect of natural and anthropogenic perturbations on the size of commercial stocks, to delineate fishery stocks and management units of living resources, to determine predator-prey relationships, and to restore habitats essential to robust fisheries.

Defining the Impact of the Seas on the Global Environment

Powerful tools are being developed to elucidate the many biogeochemical cycles that determine the fate of all the life- supporting elements on Earth. Scientists are beginning to understand and manipulate the molecular genetics and biology of esoteric metabolic pathways associated with the carbon, sulfur, phosphorous, iron, and other biogeochemical cycles. For example, based on the hypothesis that iron controls photosynthesis in the oceans, immunological probes were used to show that addition of iron to open ocean water off the Galapagos Islands significantly increased energy production.

Marine biotechnology will be useful in assessing the role of the oceans in affecting climate change and the global carbon cycle. Molecular techniques can facilitate and enhance the measurements of CO_2 concentrations and total CO_2 inventories being developed for global ocean models of carbon cycling. There is compelling evidence that the exchange of dissolved and particulate materials between the continental shelf and its boundaries is a significant factor affecting the flux of CO_2 and biogenic elements within the global ocean. Several Federal agencies plan to collaborate on related research, including marine biotechnology applications.

Fish Health Management: Overview

When working with fish, disease prevention is always more rewarding than treatment. Once fish are sick, accurately identifying all problems present can be difficult, and treatment must be administered early in the course of an epizootic to be effective. In most cases, a comprehensive program of fish health management should be based on the principles of water quality, nutrition, sanitation, and quarantine. Water quality and nutrition are discussed below. Sanitation includes maintenance of a clean environment with minimal accumulation of organic debris, proper disinfection of nets and equipment, and thorough disinfection of fish-holding units between groups of fish. While these practices are more applicable to hatchery management and intensive, indoor systems than to pond or sea-cage culture, efforts to maintain as clean an environment as possible within the constraints of the operation will help minimize disease outbreaks. New fish should be quarantined for at least 3 wk, with 30-60 days preferred for valuable pets or zoologic specimens. Producers may be constrained by production practices; however, they may be forced to address quarantine concerns as regulations are enforced in an effort to prevent the introduction of viral diseases to aquaculture facilities.

Aquatic medicine has emerged as a recognised specialty within the practice of zoological medicine. Fish medicine, an important component of the aquatic specialty, is evolving with distinct subspecialties of aquaculture or production medicine, and pet fish medicine that focuses on individual animals.

Although reference will be made to aquaculture medicine, the emphasis of this chapter is on pet fish medicine. In addition to aquaculture and pet fish practice, aquatic species are increasingly important as laboratory models for biomedical research and toxicology studies. Maintenance and husbandry of aquatic species is becoming a component of laboratory animal practice.

Many fish sold through the pet trade are raised on farms in south Florida. Live bearers (ie, guppies, platties, and swordtails) are pond raised and harvested by trapping. Most other fish are categorized as egg layers, which are often hatched indoors and then moved into fertilized ponds for grow-out. At the end of the production cycle, which is usually 3-6 mo, ponds are drained, and fish are harvested with a seine net and prepared for shipment. A large percentage of fish sold through the pet trade is imported.

Some of these fish are wild caught, while others are raised on small production facilities, many of which are located in southeast Asia. Virtually all marine fish, except for clown fish (Amphiprion spp), are wild caught. The pet trade is a global industry, and fish may be moved through several dealers before reaching a retail outlet.

Koi and fancy goldfish for ornamental garden ponds have grown in popularity. Many are imported from Japan (koi) or China (fancy goldfish). Imported, show quality fish sell for hundreds to thousands of dollars each. Koi are very hardy and quite tolerant of medical and surgical procedures. Owners value individual animals and are often eager to invest in veterinary care for these pets. Clinical management of individual pet fish, exhibit animals, and valuable broodstock has changed dramatically in recent years.

Advances include use of nonlethal methods for diagnosing disease and more sophisticated treatment options. Radiology and ultrasound are particularly well suited for disease diagnosis in aquatic species. Development of blood culture techniques to accurately identify bacterial agents and run sensitivity tests prior to the start of antibiotic therapy has been particularly useful in decreasing the need to euthanize, or surgically biopsy, an animal to achieve an accurate diagnosis. Surgical advances, including use of exploratory laparotomy and swim bladder repairs, have salvaged animals that previously would have been euthanized.

Physiology

Fish are poikilothermic, and all physiologic processes are greatly influenced by water temperature. In freshwater, the internal tissues of fish are hyperosmotic, whereas in saltwater they are hypo-osmotic. Surface injuries to the skin make osmoregulation more difficult and may be of serious consequence due to loss of fluid balance and circulatory collapse.

Fish lack organised lymph nodes and Kuppfer's cells. Phagocytic tissue is located in the hematopoietic tissue of the spleen and kidney and often in the atrium of the 2-chambered heart. The structure of the fish kidney varies with the species; generally it is divided into an anterior "head" kidney and a posterior "caudal" kidney, located retroperitoneally, ventral to the vertebral column. Hematopoietic, renal, and endocrine tissues are found in the kidney. Divalent ions are excreted principally via the kidney, and monovalent ions and nitrogenous excretions via the gills. Accordingly, lesions of the kidney and gills may seriously interfere with respiration, excretion, and fluid balance.

The swim bladder in bony fish, which originates as an appendage of the foregut, regulates body buoyancy and may also be used for sound production. Gas is either secreted by or absorbed into the swim bladder to maintain buoyancy or specific gravity and balance. A sensory lateral line system along the sides of the body and head receives stimuli from the aquatic environment and mediates adaptive responses through the CNS.

A humoral antibody system occurs in all fish but varies considerably between classes. Although antibody production often is temperature dependent, specific serum antibodies can be demonstrated. B lymphocytes, found in the spleen and liver, are responsible for production of immunoglobulins found in the serum and tissue fluids of fish. However, fish lack the potent immunoglobulins similar to IgG of other animals. Fish do increase production of IgM, similar to other vertebrates, when responding immunologically to many infectious agents. Fish depend on increases in environmental temperature for efficient antibody production during infections (or after vaccinations), when most pathogens are replicating at a more rapid rate. The optimal temperature for antibody production varies with the species of fish (warmwater or coldwater). Extreme increases in environmental temperature (above that of the natural habitat) inhibit antibody production. T lymphocytes of fish, like those of other vertebrates, are responsible for cell-mediated immunity. Immunity is not as age dependent in fish as it is in other animals; young fish are usually immunocompetent and can be vaccinated successfully. Antibodies are found in the mucus of the fish skin and GI tract.

While anamnestic immunologic responses have been documented in fish, the duration of acquired immunity appears limited. Immunity lasts longer with individual parenteral administration of antigens than with mass bath methods. Although vaccination of fish against specific diseases has been economically important in preventing losses, there is a need for improved methodology.

2

Biotechnology in Fish Breeding

Gonadotropin releasing hormone (GnRH) is now the best available biotechnological tool for the induced breeding of fish. GnRH is the key regulator and central initiator of reproductive cascade in all vertebrates (Bhattacharya et al.,2002). It is a decapeptide and was first isolated from pig and ship hypothalami with the ability to induce pituitary release of luteinising hormone (LH) and follicle stimulating hormone (FSH) (Schally et al.,1973). Since then only one form of GnRH has been identified in most placental mammals including human beings as the sole neuropeptide causing the release of LH and FSH. However ,in non mammalian species (except guinea pig) twelve GnRH variants have now been structurally elucidated ,among them seven or eight different forms have been isolated from fish species.(Halder et al.,1991;Sherwood et al.,1993;King and Miller,1995;Jimenez-Linan et al.,1997). The most recent GnRH purified and characterized was by Carolsfeld et al.(2000) and Robinson et al.(2000). Depending on the structural variant and their biological activities, number of chemical analogues have seen prepared and one of them is salmon GnRH analogue profusely used now in fish breeding and marked commercially under the name of ýÿOvaprimýÿ throughout the world .The induced breeding of fish is now successfully achieved by development of GnRH technology.

Transgenesis

Transgenesis or transgenics may be defined as the introduction of exogenous gene / DNA into host genome resulting in its stable maintenance, transmission and expression. The technology offers an excellent opportunity for modifying or improving the genetic traits of commercially important fishers, mollusks and crustaceans for aquaculture. The idea of producting transgenic animals became popular when Palmitter et al. (1982) first produced transgenic mouse by introducing metallothionein human growth hormone fusion gene (mT-hGH) into mouse egg, resulting in dramatic increase in growth. This triggered a series of attemptson gene transfer in economically important animals including fish.

The first transgenic fish was produced Zhu et al. (1985) in China, who claimed the transient expression n putative transgenics, although they gave no molecular evidence for the integration of the transgene. The technique has now seen successfully applied to a number of fish species. Dramatic growth enhancement has been shown using this technique especially in salmonids (Devlin et al., 1994). Some studies have revealed enhancement of growth in adult salmon to an average of 3 ýÿ 5 times the size of non ýÿ transgenic controls, with some individuals, especially during the first few months of growth, reaching as much as 10 ýÿ 30 times the size of the controls (Devlin et al., 1994; Hew et al., 1995).

The introduction of transgenic technique has simultaneously put more emphasis on the need for production of sterile progeny in order to minimize the risk of transgenic stocks mixing in the wild populations. The technical development has expanded the possibilities for producing either sterile fish or those whose reproductive activity can be specifically turned on or off using inducible promoters.

This would clearly be of considerable value allowing both optimal growth and controlled reproduction of the transgenic stocks while ensuring that any escaped fish would be unable to breed. An increased resistance of fish to cold temperatures has been another subject of research in fish transgenics for the past several years (Fletcher et al., 2001). Coldwater temperatures pose a considerable stressor to many fish and few are able to survive water temperatures much below 0-1°C. this is often a major problem in aquaculture in cold climates. Interestingly, some marine teleosts have high levels (10 ýÿ 25 mg/ml) of serum antifreeze proteins (AFP) or glycoproteins (AFGP) which effectively reduce the freezing temperature by preventing ice-crystal growth.

The isolation, characterization and regulation of these antifreeze proteins particularly of the inter flounder Pleuronectas americanus has been the subject of research for a considerable period in Canada. Consequently, the gene encoding the liver AFP from winter flounder was successfully introduced into the genome of Atlantic salmon where it became integrated into the germ line and then passed onto the off ýÿ spring F3 where it was expressed specifically in the liver (Hew et al., 1995).

The introduction of AFPs to gold fish also increased their cold tolerance, to temperatures at which all the control fish died (12 h at 0o C; Wang et al., 1995). Similarly, injection or oral administration of AFP to juvenile milkfish or tilapia led to an increase in resistance to a 26 to 13o C. drop in temperature (Wu et al., 1998). The development of stocks harbouring this gene would be a major benefit in commercial aquaculture in counties where winter temperatures often border the physiological limits of these species.

The most promising tool for the future of transgenic fish production is undoubtedly in the development of the embryonic stem cell (ESC) technology. There cells are undifferentiated and remain totipotent so they can be manipulated in vitro and subsequently reintroduce into early embryos where they can contribute to the germ line of the host. This would facilitate the genes to be stably introduced or deleted (Melamed et al., 2002). Although significant progress has been made in several laboratories around the world, there are numerous problems to be resolved before the successful commercialization of the transgenic brood stock for aquaculture.

To realise the full potential of the transgenic fish technology in aquaculture, several important scientific break ýÿ through are required. There include (i) more efficient technologies for mass gene transfer (ii) targeted gene transfer technologies such as embryonic stem cell gene transfer (iii) suitable promoters to direct the expression of transgenes at optimal levels during the desired developmental stages (iv) identified genes of desireable traits for aquaculture and other applications (v) informations on the physiological, nutritional, immunological and environmental factors that maximize the performance of the transgenics of the transgenics and (vi) safety and environmental impacts of transgenic fish.

Chromosome Engineering

Chromosome sex manipulation techniques to induce polyploidy (triploidy and tetraploidy) and uniparental chromosome inheritance (gynogenesis and androgenesis) have been applied extensively in cultured fish species (Pandian and Koteeswaran, 1998; Lakra and Das, 1998). These techniques are important in the improvement of fish breeding as they provide a rapid approach for gonadal sterilization, sex control improvement of hybrid viability and clonation. Most vertebrates are diploid meaning that they possess two complete chromosome sets in their somatic cells. Polyploidy individuals possess on or more additional chromosome sets, bringing the total to three in triploids, four in tetraploids and so on.

Induced triploidy is widely accepted as the most effective method for producing sterile fish for aquaculture and fisheries management. The methods used to induce triploids and other types of chromosome set manipulations in fishes and the applications of these biotechnologies to aquaculture and fisheries management are well described (Purdom, 1983; Chourrout, 1987; Thorgaard, 1983; Pandian and Koteeswan, 1998). Tetraploid breeding lines are of potential benefit to aquaculture, by providing a convenient way to produce large numbers of sterile triploid fish through simple interploidy crosses between tetraploids and diploids (Chourrout et al., 1986; Guo et al., 1996). Although tetraploidy has been induced in many finfish species, the viability of tetraploids was low in most instances (Rothbard et al., 1997).

In teleosts, technique for inducing sterility include exogenous hormone treatment (Hunter and Donaldson, 1983) and triploidy induction (Thorgaard, 1983). The use of hormone treatements, however could be limited by governmental regulation and a lack of consumer acceptance of hormone treated fish products. Triploidy can be induced by exposing eggs to physical or chemical treatment shortly after fertilization to inhibit extrusion of the second polar body (For reviews see Purdom, 1983; Thorgaard, 1983 and Ihssen et al., 1990) triploid fish are expected to be sterile because of the failure of homologous chromosomes to synapse correctly during the first meiotic division. Methods of triploidy induction in clued exposing fertilized eggs to temperature shock (hot or cold), hydrostatic pressure shock or chemicals such an colchicines, cytochalasin-B or nitrous oxide. Triploid can also be produced by crossing teraploids and diploids. Tetraploid induction involves fertilizing eggs with normal sperm and exposing the diploid sygote for physical or chemical treatment to suppress the first mitotic division. Gynogenesis is the process of animal development with exclusive maternal inheritance.

The production of gynogenetic individuals is of particular interest to fish breeders because a high level of inbreeding can be induced in single generation. Gynogenesis may also be used to produce all ýÿ female populations in species with female homogamety and to reveal the sex determination mechanisms in fish. It is convenient to use all female gynogenetic progenies (instead of normal bisexual progenies) for sex inversion experiments. Methodologies combining use of induced gynogenesis with hormonal sex inversion have been developed for several aquaculture species (Gomelsky et al., 2000). Androgenesis is the process by which would have commercial application in aquaculture. It can also be used in generating homozygous lines of fish and in the recovery of lost genotypes from the crypreserved sperms. Androgenetic individuals have been produced in a few species of cyprinids, cichlids and salmonids (Bongers et al., 1994).

Biotechnology and Fish Health Management

Disease problem area major constraint for development of aquaculture. Biotechnological tools such as molecular diagnostic methods, use of vaccines and immunostimulants are gaining popularity for improving the disease resistance in fish and shellish species world over for viral diseases, avoidance of the pathogen in very important. In this context there is a need to rapid method for detection of the pathogen. Biotechnological tools such as gene probes and polymerase chain reaction (PCR) are showing great potential in this area. Gene probes and PCR based diagnostic methods have developed for a number of pathogens affecting fish and shrimp (karunasagar ,1999). In case of finfish aquaculture, number of vaccine against bacteria and viruses have been developed. Some of these have been conventional vaccines consisting of killed microorgansism but new generation of vaccine consisting of protein subunit vaccine genetically engineered organism and DNA vaccine are currently under development.

In the vertebrate system, immunization against disease is a common strategy. However the immune system of shrimp is rather poorly developed, biotechnological tools are helpful for development of molecule, which can stimulate this immune system of shrimp. Recent studies have shown that the non specific defence system can be stimulated using, microbial product such as lipopolysacharides, peptidoglycans or glucans (itami et al 1998). Among the immunostimulants known to be effective in fish glucan and levamisole enhance phagocytic activities and specific antibody responses (Sakai, 1999).

Cryopreservation of Gametes or Gene Banking

Cryopreservation is a technique, which involve long-term preservation and storage of biological material at a very low temperature usually at -196 C ,the temperature of liquid nitrogen. It is based on the principle that very low temperature tranquilize or immobilize the physiological and biochemical activities of cell, thereby making it possible to keep them viable for very long period.

The technology of cryopreservation of fish spermatozoa (milt) has been adopted for animal husbandary . The first success in preserving fish sperm at low temperature was reported by Blaxter (1953) who fertilizes Herring (*Clupea herengus)* eggs with frozen thawed semen .The spermatozoa of almost all cultivable fish species has now been cryopreserved (Lakra 1993) . Cryopreservation overcomes problems of male maturing before female, allow selective breeding and stock improvement and enables the conservation (Harvey ,1996). One of the emerging requirement for that can be used by breeders for evolving new strains. Most of the plant varieties that has been produced are based on the gene bank collections. Aquatic gene bank however suffers from the fact that at present it is possible to cryopreserve only the male gametes of finfishes and there in no viable technique for finfish eggs and embryos.

However, the recent report on the freezing of shrimp embryos. However, the recent report on the freezing of shrimps embryos by subramoniam and newton (1993) and Diwan and kandaswami (1997) look promising. Therefore, it is essential that gene banking of cultivated and cultivable aquatic species be undertaken expeditiously.

Aquaculture Biotechnology

Biotechnology is the fusion of biology and technology and refers to the application of biological techniques in product research and development. It involves the use of molecules (synthetic or physiological), biological materials (tissues, cells, genes) or processes derived from living organisms, to create new and practical applications for agriculture, medicine, industry and society.

What are its Applications for Reproductive Control in Fish Production?

The control of seasonal production and reproduction in farm animals have become major research goals. The applications of biotechnology to fish farming and ornamental fish production are numerous and valuable in both economic (food production, aquarium trade) and environmental terms (conservation of natural biodiversity for endangered species and protection of natural biodiversity from escapee domesticated strains). With the growing demand for fish products, biotechnology can help in the development of high quality, economical produce, thereby reducing pressure on natural populations.

In terms of fish reproduction for aquaculture, this means development of methods for the production of good quality gametes, independent of season, producing progeny with specific desired characteristics (good growth rate, flesh quality etc), meeting consumer demand. This technology should also contribute to the preservation and dissemination of genetic improvements.

Biotechnology may be used to:

- Control the sex of fish (e.g. produce single-sex populations)
- Delay the age of first sexual maturation (puberty)
- Control the reproductive cycle (in terms of gametogenesis and spawning)
- Support the genetic improvement of stocks, to produce strains more resistant to disease, displaying better growth rates (and food conversion ratios), more tolerant to stress and temperature changes, providing a product with an improved nutritional value, taste, texture and appearance.
- Improve management of gametes (short and long-term storage, transportation, development of media for in-vitro fertilisation)
- Conserve endangered or economically important strains via selective breeding programmes and management of captive broodstock (spermatogonial transplantation, cryopreservation of gametes)
- Produce sterile farmed animals to prevent the impact of escapees on the local biodiversity
- Assist in providing the aquarium trade with ornamental fish

Which Methods are Currently Practiced?

In-vitro fertilisation: This is the fertilisation of fish oocytes with manually/artificially collected sperm and ovules. Since the 1990's, research has permitted the development of several media (for dilution, to prevent spermatozoa activation, to preserve gamete fertilizing capability, for in-vitro gamete maturation...), thereby the success rates of the fertilisation process.

A dilution medium for washing oocytes, Ova-fish (IMV technologies), prior to in-vitro fertilisation, is currently available, which enables an in-vitro hatching rate success of 60-70% (for rainbow trout). Ovafish solution removes debris from the eggs, such as blood, urine and mucus, which may otherwise affect fertilisation success. Another medium, Actifish (IMV technologies), has been developed to improve the fertilising capability of fish sperm. Actifish is a fertilisation medium that can be used for fresh, conserved or frozen semence. Another medium, MaturFish (IMV technologies), has also been developed to promote sperm maturation. All these media have been specifically developed for the trout, but other examples exist for numerous species, including those of marine origin.

Long and Short Term Preservation of Gametes

Cryopreservation enables the long-term preservation of sperm of individuals of interest (disease resistant phenotypes etc.), for a virtually unlimited period of time. This is useful for many reasons, for example: to develop breeding programmes independent of maturation period; to market standard quality sperm; to provide a year round supply of gametes; and to preserve a genetic stock of both commercially important strains as well as those species threatened by extinction in the wild. Research has permitted the development of several dilution mediums to protect sperm during freezing. However, freezing technology has very limited uses.

Oocytes cannot currently be frozen without losing their viability. However, research is in progress on the vernalisation of oocytes, this is very low temperature storage, without freezing, and could preserve the oocytes for several months.

Different media and temperatures have also been investigated for the short-term storage of both eggs and sperm. For example, studies have shown that trout eggs can be stored at 12°C for 3 days in coelomic fluid with no detrimental effect on egg quality.

Sex Control

The control of fish sex could be useful where one sex displays advantageous characteristics, such as larger adult size, production of high-value caviar (sturgeon), faster growth rate, or higher age at first sexual maturation (thereby avoiding degradation of flesh quality and certain bacterial infections). Sex control is currently only practiced commercially for salmonids, tilapia and hirame.

Monosex populations of the most advantageous sex may be produced either by direct sex control via steroid treatment of the fish destined for consumption (masculinisation by administration of androgens; feminisation by administration of estrogens); or by genetic control and steroid treatment of broodstock (indirect hormonal treatment, gynogenesis, androgenesis, hybridisation); or by control of external factors (temperature, density etc.).

Genetic sex regulation may be achieved by:

- Crossing sex-reversed adult broodstock (administering androgens to produce "neo" males and estrogens to produce "neo" females; a "neo" fish is one which is genetically male or female, but phenotypically of the opposite sex) with normal males or females to produce single-sex progeny. Within Europe, only indirect androgen treatment of broodstock is permitted (i.e. production of all female progeny). This has obvious environmental advantages over the hormonal treatment of whole populations of pre-differentiated fish.
- Gynogenesis, or the fertilisation of oocytes with inactivated sperm (irradiated) with normal sperm, to eliminate the female chromosomes and produce all female progeny when the genotype is XX/XY. This method is more successful than androgenesis, but not widely used commercially.
- Androgenesis, or the fertilisation of inactivated oocytes (irradiated) with normal sperm, to eliminate the female chromosomes and produce all male progeny when the genotype is ZZ/ZW. In practice this method is not very successful and therefore not used commercially.
- Hybridisation of two different species from the same genera, producing single-sex progeny. However, the resulting offspring do not always display a satisfactory growth rate.

(Note: control of external factors, such as temperature for some gonochoristic fish - which do not change sex during their lifetime, such as tilapia or seabass; and social stimuli for hermaphrodites, such as seabream - which change sex during their lifetime, is a "gentle" method for controlling sex, but it is not yet commercially viable as it is not sufficiently reliable).

Control & Synchronisation of Spawning

The spawning season may be controlled by the manipulation of species-dependent environmental factors, such as photoperiod and temperature. These methods are routinely used to produce eggs in salmonid and some marine species such as seabass, turbot and seabream.

Spawning events may also be induced and spermiation enhanced by administration of gonadotropin releasing hormone analogs (GnRHa or LHRHa), hcG (human chorionic gonadotropin) or fish pituitary extracts. For example, within Europe, a synthetic GnRH analog,

GonazonÒ (Intervet), is officially approved as an inducer of ovulation, showing good success rates in salmonids for both ovulation and subsequent fertilisation. In some other species, such as cyprinids, more potent inducers are used (hCG or Carp Pituitary Extract).

Other speciality pharmaceutical products such as OvaprimÒ have been developed, which combine a GnRH analog with domperidone, a potent inhibitor of the dopaminergic system. Fish farmers may also control production by speeding up or slowing down embryonic development by temperature and / or feeding practices.

Control of Puberty

Puberty may also be delayed by using different photoperiod regimes, for example in Atlantic salmon production. Puberty may also be completely eliminated, at least in females, by induction of sterility by triploidy. The application of antireceptor vaccinations, which have specific antagonistic effects of gametogenesis, is also in development.

Control of Fertility

The production of reproductively sterile fish (triploids) is very useful in responsible fish farm management and limits the genetic risk associated with the escape of domesticated fish into the wild. These farmed populations cannot interbreed with wild populations and therefore do not pose a threat to the natural biodiversity. In some species sterile fish reach a larger size without the complications of sexual maturity and the consequent reduction in flesh quality, which is desirable in sectors of the fish farming industry where large fish are most valuable.

Generally, triploids grow more slowly than diploids. Triploidy is also not 100% successful (98% success rate), and as it is not feasible to test individual fish to check whether or not they are triploid, there will always be a certain number of individuals within the stock that are not triploid and are thus capable of reproducing. It should also be remembered that if the aim is to produce small sized fish (e.g. up to 200g), sexual maturity is not a problem as the fish will not attain maturity before they have reached their marketable size.

All female triploid offspring may be produced by the indirect use of externally administered hormones and by pressure / temperature treatment of eggs for a few minutes, post fertilisation. Within the European Union, it is forbidden to market hormone-treated fish for human consumption. Methyltestosterone may be used to sex-reverse broodstock females (XX) to neo-males (only sperm containing the X chromosome). These neo-males (XX) are then crossed with normal females (XX).

Induction of triploidy in the resulting zygotes will produce all-female triploid offspring (XXX). This is currently practiced on rainbow trout and perch. However, current research focuses on the development of alternative, more environmentally sensitive methods that are also species specific. Female triploids are always sterile, they do not undergo gonad development; however triploid males are only functionally sterile, they develop gonads and undergo puberty, but their sperm does not give rise to viable progeny.

Development of Gene Transfer Technologies for Fish Reproduction

Gene transfer technology has been applied to aquaculture species to obtain sterile fish by preventing the expression of GnRH. Further information can be obtained from the following papers:

- Wei Hu et al. 2007. Antisense for gonadotropin-releasing hormone reduces gonadotropin synthesis and gonadal development in transgenic common carp (Cyprinus carpio). Aquaculture 271: 498-506.
- Uzbekova, S. et al. 2000. Transgenic rainbow trout expressed sGnRH-antisense RNA under the control of sGnRH promoter of Atlantic salmon. J. Mol. Endocrinol. 25 (3): 337-350.

Principal Further Research Requirements

- Optimisation of species-specific hormonal treatments, in terms of the development of rapidly metabolised molecules that do not release biologically active catabolites into the environment (e.g. possible use of recombinant fish gonadotrophins for inducing gonadal development).
- Development of alternative methods to exogenous hormones based on genetic selection, environmental variable manipulation (e.g. photoperiod and temperature) and gene transfer technology, for fertility and sexual cycle control.
- Genetic basis of reproduction, for selection criteria / gene markers for broodstock companies
- Improved knowledge on important steps in fish reproductive cycles (sex control, puberty, gametogenesis and spawning)
- Improved knowledge on the determinants of gamete quality (e.g. surrogate broodstock technology)

- Research into reducing the impact of fish farms on the environment (in terms of the genetic integrity of the wild population, the discharge of hormones and antbiotics into the environment etc.)
- In terms of cryopreservation, a better knowledge of sperm membrane biology is needed. An improved understanding is required of the mechanisms regulating tolerance to freezing (for sperm and oocytes), as well as the formulation of improved protocols.
- Conservation of genetic resources and regeneration of populations of valuable animals or endangered species (e.g. spermatogonial transplantation).
- Development of generic methods for the domestication of new species impact of Science and Technology on Indian

Fisheries Sector

Fisheries sector, a sunrise sector in India, has recorded a faster growth than that of crop and livestock sectors. The sector contributes to the livelihood of a large section of economically-underprivileged population of the country. It has been recognised as a powerful income and employment generator as it stimulates growth of a number of subsidiary industries and is a source of cheap and nutritious food besides being a foreign exchange earner. With the changing consumption pattern, emerging market forces and technological developments, it has assumed added importance in India.

It is undergoing rapid transformation and the policy support, production strategies, public investment in infrastructure, and research and extension for fisheries have significantly contributed to the increased fish production. Particularly, after the mid-1980s, the development of carp polyculture technology has completely transformed this traditional backyard activity into a booming commercial enterprise.

Fish Production in the World and in India

Approximately 50 million people worldwide depend on fishing for all or most of their family earnings, while another 150 million depend on fish processing and the fleet servicing industry. More than 10 million work on 2.5 million small-scale fishing vehicles and account for 50% of the world's catch (FAO 2001).

Fish production in the world rose from 23.50 million tonnes in 1950-51 to 140.48 million tonnes in 2003-04. Correspondingly, fish production in India has touched 6.40 million tonnes in 2003-04 from a mere 0.75 million tonnes in 1950-1951 (Table 1). The total fish production of our country stood at 6.87 million tonnes in 2006-07.

The share of India in global fish production has grown gradually from about 2.66% during the 1960s and 1970s to 4.56% in 2003-04.

Table: *Fish Production in World and India, 1950-51 to 2003-04*
(Million tonnes)

Year	World	% change	India	% change	India's share
1950-51	23.50	-	0.75	-	3.19
1960-61	43.60	85.53	1.16	54.67	2.66
1970-71	66.20	51.83	1.76	51.72	2.66
1980-81	72.30	9.21	2.44	38.64	3.37
1990-91	98.26	35.91	3.84	57.38	3.91
2000-01	129.00	32.35	5.66	47.40	4.39
2003-04	140.48	8.90	6.40	13.07	4.56
2006-07*	--	--	6.87	--	--

Source: 1. ICAR-ICLARM Project on 'Strategies and Options for Increasing and Sustaining Fisheries and Aquaculture Production to Benefit Poor Households in India, 2004 and Economic survey 2006-07, GOI

2. *Annual Report 2006-07 Provisional Estimates

Table1 shows that growth in fish production in India has been at a faster rate than in the world; mainly due to increasing contribution from inland fisheries. It can also be seen in Figures that in the pre-WTO period of 1990-91, the share of India's fisheries was 3.90, which rose to 4.56 in the post-WTO period of 2003-04. Overall, the share of developing world in the total world fish production increased from 43% in 1973 to about 73% in 1997, which has been mainly due to the increasing contribution from countries like China and India (Delgado *et al*, 2003).

Contribution to Indian Economy and Prospects of Fisheries Growth

With fisheries sector comprising marine fisheries, freshwater and brackish water aquaculture and inland fisheries consisting of tanks ,and reservoir, the potentiality of this sector as a whole remains to be fully tapped and it remains a sector of much promise. The fisheries sector in particular is more complex enterprise that functions under integrated network of natural resources, other enterprises with forward and backward linkages with fisheries and other socio-political variables. The major functions of fisheries enterprises, viz. production, transportation, storage and processing involve value addition from labour, capital and management, which significantly influence the rapid economic development of the country.

In the last 25 years, unlike agriculture, the contribution of fisheries sector to GDP continued to grow at a rapid pace because of expansion of culture fisheries enterprise. The share of agriculture and allied activities in the total GDP is constantly declining. It was 34.69% in 1980-81 and declined gradually to 17.62% in 2004-05. In contrast, the contribution of fisheries sector to the total GDP has gone up from 0.75% in 1980-81 to 1.04 in 2004-05 (at current prices).

Similarly, the share of fisheries in agriculture GDP (AgGDP) has increased robustly from 2.17% in 1980-81 to 5.93% in 2004-05. This sector is in fact pushing the agricultural growth upward for the past 5 and half decades.

Table: *Comparison and Growth of Fisheries Sector*

Period	Percent contribution of		
	Agriculture to Total GDP	Fish to Total GDP	Fish to Ag GDP
1980-81	34.69	0.75	2.17
1990-91	28.42	0.96	3.37
2000-01	22.26	1.18	5.32
2004-05	17.62	1.04	5.93

Source: National Account Statistics, CSO, GOI.

3

Vertebrate-Derived Compounds

Squalamine

Squalamine is an aminosterol isolated from the stomach and liver of the spiny dogfish, *Squalus acanthus*, a common New England coastal shark species. (When it was discovered in 1993, the compound was reported to exhibit broad-spectrum antibiotic activity. Squalamine was licensed to Magainin Pharmaceuticals (now Genaera Corporation) for development. Of considerable interest is published evidence suggesting that squalamine exhibits anti-angiogenic activity under certain conditions (angiogenesis is the formation and differentiation of blood vessels).

Neovastat" (AE-941)

Neovastat (AE-941) is a derivative of shark cartilage extract. Rather than being a specific monomolecular compound, AE-941 is a defined standardized liquid extract comprising the < 500 kDa (kilodaltons, a unit of mass) fraction from shark cartilage. It has the anti-angiogenic and antitumor activity. The drug's anti-angiogenic bioactivity suggests it could be a valuable agent for use in patients suffering from multiple myeloma and other hematologic (blood) diseases.

Sponge-Derived Compounds

Bengamides and Derivatives

Two novel seven-membered ring heterocycles, bengamide A and bengamide B, were reported in 1986 isolated from an as yet undescribed Fijian sponge belonging to family Jaspidae. Since this time, a number of additional compounds from the bengamide class have been isolated, most notably from the Fijian sponge *Jaspis*cf. *coriacea*. Bengamides A and B were initially reported to exhibit *in vitro* toxicity to larynx epithelial carcinoma cells, and to have antibiotic and anti-helminthic activity (against the nematode *Nippostrongulus braziliensis*).

Contignasterol (IZP-94005, IPL576,092)

Contignasterol (IZP-94005) from the sponge *Petrosia contignata* was first reported in the early 1990s. The natural compound itself was deemed a good anti-asthma drug candidate based on performance in vivo animal models. One contignasterol derivative, named IPL576,092, shows promise as an oral asthma medication. More recently, IPL576,092 has entered clinical trials as a treatment for diseases causing inflammation of the eyes and skin.

Debromohymenialdisine (DBH)

Debromohymenialdisine (DBH) is an alkaloid originally isolated from the shallow-water Palauan marine sponge *Stylotella aurantium*. The molecule is intriguing not just for its drug-like properties, but also because it's simple molecular structure has yielded to easy total synthesis in the laboratory. Research suggest the compound acts as a highly selective inhibitor of a specific target cell DNA damage checkpoint enzyme during the G2 phase of the cell cycle. The compound is a promising potential Anti-Alzheimer agent.

Girolline (Girodazole)

Girolline was reported to inhibit protein synthesis in eukaryotic target cells. The compound was of interest because it inhibited protein synthesis at the termination steps of the process rather than at the initiation or chain elongation steps. The simple structure of the compound facilitated the production of a modified synthetic analog.

Halichondrins

A compound given the name halichondrin B, belonging to a chemical family known as the macrolides, was isolated by Uemura *et al.* (1985) from the Japanese sponge *Halichondria okadai*. Initial investigations into the bioactivity of the compound revealed that halichondrin B apparently bound tubilin at a site close to the so-called vinca site and altered tubulin depolymerization.

Hemiasterlins (H-286)

This class of novel marine oligopeptides have been shown to be potent antitumor agents. Representatives of this class of natural products have been isolated from extracts prepared from sponges residing in two distinct genera (*Auletta; Siph-onochalina*). Three different hemiasterlins with drug develop-ment potential (hemiasterlin, hemiasterlin A, hemiasterlin C) have been the subject of chemical and biological investiga-tions. These molecules exhibit cytotoxic and antitubulin activity similar to that seen in the dolastatins. Mitotic inhibition occurs through binding to tubulin at the vinca/peptide region in a manner similar to dolastatin.

KRN7000

It was first isolated from the sponge *Agelas mauritianus*. These compounds, dubbed the agelasphins, were shown to exhibit antitumor and possible immunostimulatory activity.

Lasonolides

The lasonolides are a series of marine natural products under investigation for the treatment of cancer. The compounds were isolated in 1994 by scientists from the Harbor Branch Division of Biomedical Marine Research from the sponge *Forcepia* sp. found in Gulf of Mexico deep-sea habitats. These compounds are very potent and show especially promising properties for the treatment of pancreatic cancer. They kill cancer cells in a different way than most other cancer drugs. The exact mode of action is not yet fully understood, and is an area of active research. In addition to anti-proliferative and antitumor properties, these novel macrolide compounds display antifungal activity as well.

Manoalide

Manoalide was isolated from the sponge *Luffariella variabilis* collected in the Indo-Pacific. It is a member of a chemical family known as the sesquiterpenes. (Although this natural product was originally reported as an antibiotic, follow-up work revealed manoalide possesses promising analgesic and anti-inflammatory properties. The compound works by inhibiting Phospholipase A2 (PLA2), which plays an important role in the inflammation process.

Topsentins

The Topsentins are a class of natural products that have been extracted from several sponge species subsequent to their initial isolation from the sponge *Spongosorites ruetzleri*. These compounds have been shown to have significant anti-inflammatory properties. Although the precise mode of action is not known, the compound has been reported capable of suppressing immunogenic as well as neurogenic (originating in nerve tissue) inflammation. The topsentins may hold promise as an arthritis medication or as additives in anti-inflammatory creams for the treatment of skin irritations.

Dictyostatin

Dictyostatin was first isolated in 1994 by Bob Pettit and his colleagues at Arizona State University from a sample of an unidntified dictyoceratid sponge of genus *Spongia* collected in Jamaica. The compound was later isolated from a lithistid sponge of the family Corallistidae and has been more fully repurified and characterized by researchers at Harbor

Branch Oceanographic Institution and elsewhere. Dictyostatin inhibits the growth of human cancer cells and has been shown active against certain Taxol-resistant tumors. Its mechanism of action appears to be prevention of the breakdown of tubulin during mitosis in a fashion similar to the successful cancer drug Taxol.

Laulimalide

Sometimes referred to as the fijianolides, the natural products laulimide and (the significantly less bioactive) isolaulimide were first extracted from the Pacific sponge *Cacospongia mycofijiensis*. The bioactivity displayed by these compounds is as microtubule stabilizing agents potentially arresting the development of target cells.

Cnidarian Derived Compounds

Pseudopterosins The pseudopterosins were isolated from a Caribbean soft coral species called a sea whip (Pseudopt-erogorgia elisabethae). They belong to a class of compounds known as tricyclic diterpene glycosides. Pseudopterosins have been shown to possess potent anti-inflammatory and analgesic (pain relief) properties. They appear to work by inhibiting the synthesis of eicosanoids, (locally functioning hormone-like substances) in specific white blood cells called polymorphonu-clear leukocytes.

Eleutherobin

Eleutherobin was first found in extracts made from the octocoral *Eleutherobia* sp. collected in Australia. The reported bioactivity of this natural product, currently under preclinical investigation, is as a microtubule binding agent similar to the anti-cancer drug taxol.

Sarcodictyins

Sarcodictyins were isolated from two Mediterranean coral species, *Sarcodictyon roseum* and *Eleutherobia aurea.* It acts as tubulin interactive agents.

Helminth Derived Compounds

Anabaseine (Hoplonemertine Toxin)

Anabaseine is a nicotinoid alkaloid. It is capable of stimulating vertebrate neuromuscular nicotinic receptors and increasing cholinergenic transmission. As such it has potential as a treatment of cognitive function loss. A synthetic analog, DMXBA (GTS-21) has exhibited memory enhancing effects in recipients. The compound is currently under license by the Japanese pharmaceutical company Taiho and is in Phase I trials for treating Alzheimer's disease.

Microbe Derived Compounds

Cryptophycins

This family of compounds was initially reported from a terrestrial source (the *Nostoc* sp. cyanophyte component of a Scottish lichen). Representative compounds have also been found in species of free-living marine and non-marine cyanophytes and, more recently, from the Japanese sponge *Dysidea arenaria.* The expressed bioactivity that was originally pursued was activity as an antifungal agent. Development of the compound for this purpose was not pursued beyond preliminary investigations when it was deemed too toxic for human use. More recently it has been reported that cryptophycin 1 is an inhibitor of tubulin assembly in cells.

Curacin A

First isolated from *Lyngbya majuscula* by Gerwick et al. and reported in 1994, this natural product appeared to be a very potent tubulin interactive compound. It proved to be highly insoluble, however, so much so that bioactivity could not be demonstrated with *in vivo* animal models. These compounds are currently undergoing preclinical evaluation as potential future drugs.

Thiocoraline

Belonging to the chemical family known as the thiodepsi-peptides, thiocoraline was first isolated from *Micromonospora marina,* an actinomycete bacterium collected from coastal Mozambique, southeast Africa. The Spanish marine drug company PharmaMar has reported that the compound shows activity against several standard drug screens, including breast cancer, colon cancer, renal cancer, and melanoma. Target cells appear to be inhibited through inhibition of DNA polymerase ýÿ enzyme. The most recent published literature suggests thiocoraline is still undergoing advanced preclinical evaluation.

The Future of Marine Bioprospecting

In the near future, marine bioprospecting efforts will likely focus not only on natural products from ocean plants, animals, and microbes, but also on the potential for biotech to exploit the information stored in the genomes of these organisms. In the wake of the Human Genome Project and with the expertise and technology that emerged from it, scientists are sequencing increasing numbers non-human genomes. Total genome sequencing of a handful of marine microbes has been completed, and the first sequencing of marine vertebrate (e.g., fish) genomes has commenced as well. This and similar research will likely lead to the development of gene probes that can identify the genes in target organisms that code for the elaboration of novel natural products of potential value to humans. Increasingly, new projects are expected to target the as yet unrealised biomolecular potential of the vast and almost entirely unknown marine microbial community. These research efforts could be instrumental in developing the next generation of pharmaceuticals for improving human health, as well as contributing to improved animal and agricultural crop health.

In summary, the marine world has become an important source of therapeutic agents with novel mechanisms of action. Even though thousands of new molecules are discovered every year only small number of candidates is incorporated in clinical trials. The main problem underlying this is sustainable supply of these compounds from natural sources. To battle this problem various strategies are developed, such as mariculture or aquaculture of source organisms, development of synthetic analogues of active compounds, fermentation of microorganisms producing the compound, etc. Another possible solution is the use of genetic engineering to transfer the genes encoding the synthetic enzymes that produce the desired compound to microorganisms that can be grown in huge quantities. Development of these products and services, as well as the fundamental research from which they must be derived will be enhanced by greater dependence on interdisciplinary sciences such as pharmacology, chemical ecology, molecular biology, genomics, metagenomics, computational and combinatorial chemistry and biology.

The field of marine natural products is passing its discovery phase and moving to the second phase where understanding relationships and processes is driving the research towards novel drugs from the sea. Marine plants, animals and microorganisms will be the basis of new products and services important to technology in the future. With rich biodiversity and vast marine resources along the Indian coast, in the form of estuaries, creeks, deep seas and continental shelf, the opportunities for research in the area of marine drug development are endless.

Probiotics and its Applications in Aquaculture

India is blessed with enormous aquatic resources. In fact in a country like India, providing sustainable livelihood opportunities and food security for ever increasing rural population is perhaps the greatest national challenge, where aquaculture plays a pivotal role in meeting this challenge. As per UNO projection on population over the next 50 years, India will exceed China as the most populous country in the world and taking this alarming situation, we should assume agriculture along with aquaculture and livestock as the most appropriate farm activities and as a tool for poverty reduction, food security and overall rural development. So blue revolution can be a solution to feed this ever increasing population. Although aquaculture in India has shown a rapid progress during past few years but some major problems are hindering the progress path and disease being one of them. Hence for successful aquaculture we must ensure quality feed, good environment and disease free seeds and juveniles.

With intensification of aquaculture involving high stocking densities coupled with external inputs of feed and fertilizer, sometimes of doubtful quality, the fish farmer is now face to face with diseases of various kinds from protozoan to helminth parasites to fungal, bacterial and viral diseases and syndromes. This has resulted in increased use of chemicals and antibiotics that are anyway not a welcome addition to the human food through this important food chain. India suffered the worst disaster in mid eighties due to epizootic ulcerative syndrome (EUS) which lead to heavy mortality of fishes in ponds, tanks, rivers and reservoirs.

The country again suffered a major setback again in brackish water aquaculture when the most valuable shrimp crops meant for export were lost in almost all maritime states particularly Andhra Pradesh. This was due to white spot syndrome and due to this a large number of farms were closed and farmers gave up shrimp farming which lead to decline in foreign exchange through shrimp farming. Fish and shellfish management have to be given their dues if the industry has to be sustainable and the farmers and entrepreneurs have to be making profits. Antibiotics have sometimes been used to reduce disease; however indiscriminate use has in some cases led to increased antibiotic resistance and problem of tissue residues and trade issues. There is a wide variety of vaccines available for other sectors but a very few in fisheries sector. The reason is very little research in fish pathology particularly in our country. So a question comes how to control disease? However, one measure that might be of assistance is the use of "probiotics". "Probiotics" generally includes bacteria, cyanobacteria, micro algae fungi, etc. Some Chinese researchers translate it into English as "Normal micro biota" or "Effective micro biota"; it includes Photosynthetic bacteria, *Lactobacillus, Actinomycetes, Nitrobacteria, Denitrifying* bacteria, Bifidobacterium, yeast, etc. Usually, it does not include micro algae. In English literature, probiotic bacteria are generally called the bacteria which can improve the water quality of aquaculture, and (or) inhibit the pathogens in water there by increasing production.

"Probiotics", "Probiont", "Probiotic bacteria" or "Beneficial bacteria" are the terms synonymously used for probiotic bacteria. Probiotics are a cultured product or live microbial feed supplement, which beneficially affects the host by improving its intestinal balance and health of the host. The first probiotic discovered long time ago was *Lactobacillus* sp., the lactic acid producing bacteria.

Current Research Status of Probiotics

Nogami and Maeda (1992) isolated a bacteria strain from a crustacean culture pond. The bacterial strain was found to improve the growth of crab (*Portunus trituberculatus*) larvae and repress the growth of other pathogenic bacteria, especially *Vibrio spp.,* but would not kill or inhibit useful micro algae in sea water when it was added into the culture water. Among the bacteria population present in the culture water of the crab larvae, the numbers of *Vibrio spp.* and pigment bacteria decreased or even became undetectable when the bacteria was added into culture water. The production and survival rate of crab larvae were greatly increased by the addition of the probiotic bacteria into the culture water. They also suggested that the bacterium might improve the physiological state of the crab larvae by serving as a nutrient source during its growth. This bacterium may have a good effect in the crab larval culture as a biocontrolling agent in the future.

Austin et al (1992) reported a kind of micro algae (*Tetraselmis suecica*), which can inhibit pathogenic bacteria of fish. *Teraselmis suecica* was observed to inhibit *Aeromonos hydrophila, A. salmonicida, Serrstia liquefaciens, Vibrio anguillaram, V. salmonicida* and *Yersnia ruckeri* type I. When used as a food supplement, the algal cells inhibited laboratory-induced infection in Atlantic salmon.

When used therapeutically, the algal cells and their extracts reduced mortalities caused by *A. salmonicida, A. liquefaciens, V. anguillaram, V. salmonicida* and *Yersnia ruckeri* type I. They suggested that there may be some bioactive compounds in the algal cells, and there appears to be a significant role for Tetraselmis in the control of fish diseases.

Smith and Davey (1993) reported that a fluorescent strain pseudomonad bacteria can competitively inhibit the growth of fish pathogen *A. salmonicida.*

Their results show that the fluorescent pseudomonad is capable of inhibiting the growth of *A. salmonicida* in culture media and that this inhibition is probably due to competition for free iron. In a challenge test of the Atlantic salmon by *A. salmonicida,* a statistically significant reduction in the frequency of stress-induced infection in the group of fish bathed in the bacterium fluorescent pseudomonad compared to the control group was observed.

Austin et al (1995) reported a probiotic strain of *Vibrio alginolyticus*, which did not cause any harmful effect in salmonids. By using the cross-streaking method, the probiont was observed to inhibit the fish pathogens. When the freeze-dried culture supernatant was added to the pathogenic bacteria such as *V. ordalii, V. anguillarum, A. salmonicida* and *Y. ruckeri*, showed a rapid or steady decline in the number of culturable cells, compared to the controls. Their results indicated that application of the probiont to Atlantic salmon culture led to a reduction in mortalities when challenged with *A. salmonicida* and to a lesser extent *V. anguillarum* and *V. ordalii*. The observation with this probiotic *Vibrio* is encouraging, and it appears that there is tremendous potential for the use of such probiotics in aquaculture as part of a disease control strategy Maeda and Nagami (1989) reported some aspects of the biocontrolling method in aquaculture. In their study bacterial strains possessing vibrio static activity which improved the growth of prawn and crab larvae were observed. By applying these bacteria in aquaculture, a biological equilibrium between competing beneficial and deleterious microorganisms was produced, and results show that the population of *Vibrio spp.*, which frequently causes large scale damage to the larval production, was decreased. Survival rate of the crustacean larvae in these experiments showed much higher than those without the addition of bacterial strains. They hope that addition of these strains of bacteria will repress the growth of *Vibrio spp.*, fungi and other pathogenic microorganisms. Their data suggest that controlling the aquaculture ecosystem using bacteria and protozoa is quite possible and if this system is adopted, it will maintain the aquaculture environment in better condition, which will increase the production of fish and crustaceans.

Garriques and Arevalo (1995) reported that the use of *V. alginolyticus* as a probiotic agent may increase survival and growth in *P. vannamei* postlarvae by competitively excluding potential pathogenic bacteria, and can effectively reduce or eliminate the need for antibiotic prophylaxis in intensive larvae culture system. They believe that in nature a very small percentage of *Vibrio sp.* is truly pathogenic, and the addition of potentially pathogenic bacteria to aquaculture system through water, algae, and/or Artemia was recognised. In their study, the addition of the bacteria *V. alginolyticus* as a probiotic to mass larvae culture tanks resulted in increased survival rates and growth over the controls and the antibiotic prophylaxes.

Jiravanichpaisal and Chuaychuwong *et al* (1997) reported the use of *Lactobacillus sp.* as the probiotic bacteria in the giant tiger shrimp (*P. monodon* Fabricius). They designed to investigate an effective treatment of *Lactobacillus sp.* against vibriosis and white spot diseases in *P. monodon.* They investigated the growth of some probiotic bacteria, and their survival in the 20 ppt sea water for at least 7 days. Inhibiting activity of two *Lactobacillus* sp. against *Vibrio sp.*, *E. coli*, *Staphylococcus sp.* and *Bacillus subtilis* was determined.

Direkbusarakom and Yoshimizu *et al* (1997) reported *Vibrio spp.* which dominate in shrimp hatchery against some fish pathogens. Two isolates of *Vibrio spp.* which are the dominant composition of the flora in shrimp hatchery, were studied for antiviral activity against infectious haematopoietic necrosis virus (IHNV) and Oncorhynchus masou virus (OMV). Both strains of bacteria showed the antiviral activities against IHNV and OMV by reducing the number of plaque. Their results demonstrate the possibility of using the *Vibrio* flora against the pathogenic viruses in shrimp culture.

Sugita and Shibuga (1996) reported the antibacterial abilities of intestinal bacteria in freshwater cultured fish. They isolated bacteria from the intestine of 7 kinds of freshwater cultured fish, and investigated the antibacterial abilities of these bacteria to 18 fish or human common pathogenic bacteria. Their results indicated that the bacteria isolated from intestine of 7 kinds of freshwater cultured fish possess the antibacterial abilities, and the presence of the intestinal bacteria can protect the fish against the infection by pathogenic bacteria.

Maeda and Liao (1992) reported on the effect of bacterial strains obtained from soil extracts on the growth of prawn larvae of *P. monodon*. Higher survival and molt rates of prawn larvae were observed in the experiment treated with soil extract, and the bacterial strain which promoted the growth of prawn larvae was isolated. They have assumed that if a specific bacterium is cultured and added to the prawn ecosystem to the level of 10 million cell/ml, other bacteria may hardly inhibit the same biotype because of protozoan activity which shall be one of the way to biologically control the aquaculture water biotype and ecosystem.

Maeda and Nogami *et al* (1992) have reported the utility of microbial food assemblages in culturing a crab, *Portunus trituberculatus*. Assemblages of microorganisms were produced by adding several nutrients, urea, glucose and potassium phosphate, to natural seawater with gentle aeration in which bacteria and yeast were prevailing. When these cultured microbes were added to sea water where crab larvae of *Portunus trituberculatus* were reared, bacteria numbers decreased very rapidly, followed by the decrease in flagellated protozoa and diatoms. Their results suggest that the crab larvae fed on these microorganisms successively. They found some strains of bacteria promoted larval growth, although yeasts did not support its growth. By adopting these assemblages of microorganisms a high yield was obtained for a prawn larva *P. japonicus*, although the success was not always consistent.

Douillet and Langdon (1994) have reported use of probiotics for the culture of larvae of the Pacific oyster (*Crassostrea gigas* Thunbeerg). They added probiotic bacteria as a food supplement to xenic larval cultures of the oyster *Crassostrea gigas* which consistently enhanced growth of larvae during different seasons of the year. Probiotic bacteria were added, at 0.1 million cells/ml, to cultures of algal-fed larvae, the proportion of larvae that are set to produce spat, and subsequently the number of spat increased. Manipulation of bacterial population present in bivalve larval cultures is a potentially useful strategy for the enhancement of oyster production. They suggest that the mechanisms of the action of probiotic bacteria are providing essential nutrients that are not present in the algal diets or improving the oyster's digestion by supplying digestive enzymes to the larvae or removing metabolic substances released by bivalves or algae.

Maeda and Liao (1994) have reported microbial processes in aquaculture environment and their importance in increasing crustacean production. They suggested that based on the photosynthesis of micro algae mainly, it was clarified that bacteria, protozoa and other microorganisms from microbial food assemblages use the organic matter produced by the algae and that these assemblages play a significant role in the aquatic food chain. The growth of the larvae and their production were markedly promoted by the probiotic bacteria. In their paper, they also described the presence of a bacterial clump, stained with a fluorescent dye, inside the digestive organ of the crab *Portunus trituberculatus.*

In China, the studies on probiotics in aquaculture were focused on the photosynthetic bacteria. Qiao Zhenguo et al (1992) have studied three strains of photosynthetic bacteria used in prawn (*P. chinensis*) diet preparation and their effect. Addition of the photosynthetic bacteria in the food or culture water was found to improve the growth of the prawn and the quality of the water. Cui Jingjin et al (1997) have reported on the application of photosynthetic bacteria in the hatchery rearing of *P. chinensis.* They used a mixture of several kinds of photosynthetic bacteria (*Rhodomonas sp.*) as water cleaner and auxiliary food.

Their results showed that the water quality of the pond treated with the bacteria was remarkably improved, the fouling on the shell of the larvae was reduced, the metamorphosis time of the larvae was 1 day or even earlier, and the production of post-larvae was more than that of the control.

Wang Xianghong *et al* (1997) have done some research work on probiotic bacteria in shrimp aquaculture. On the basis of studies on intestinal micro flora of wild adult shrimp *P. chinensis,* they have chosen some probiotic bacteria from shrimp intestinal flora. When the two probiotic bacterial strains were added to the larval culture water, the survival rate, the abilities of disease resistence and low salinity tolerance were improved; average body length and weight were increased. In addition, the probiotic bacteria, when added to the larval culture water was found not to influence the total bacterial number and water quality of the sea water. They also found that some probiotic bacteria can produce some digestive enzymes; these enzymes may improve the digestion of shrimp larvae, thus enhancing the ability of stress resistance and health of the larvae.

Mechanism of Action of the Probiotic Bacteria

The mechanism of action of the probiotic bacteria has not been studied systematically. According to some recent publications, in the aquaculture the mechanism of action of the probiotic bacteria may have several aspects.

1. Probiotic bacteria may competitively exclude the pathogenic bacteria or produce substances that inhibit the growth of the pathogenic bacteria.
2. Provide essential nutrients to enhance the nutrition of the cultured animals.
3. Provide digestive enzymes to enhance the digestion of the cultured animals.
4. Probiotic bacteria directly uptake or decompose the organic matter or toxic material in the water improving the quality of the water.

Chinese researchers have done some studies on the probiotic bacteria to improve the shrimp culture water, and achieved remarkable results (Li Zhuojia et al 1997). For example, when photosynthetic bacteria was added into the water, it could eliminate the NH3-N, H2S and organic acids, and other harmful materials rapidly, improve the water quality and balance the pH.

The heterotrophic probiotic bacteria may have chemical actions such as oxidation, ammoniafication, nitrification, denitrification, sulphurication and nitrogen fixation. When these bacteria were added into the water, they could decompose the excreta of fish or prawns, remaining food materials, remains of the plankton and other organic materials to CO2, nitrate and phosphate.

These inorganic salts provide the nutrition for the growth of micro algae, while the bacteria grow rapidly and become the dominant group in the water, inhibiting the growth of the pathogenic microorganisms. The photosynthesis of the micro algae provide dissolved oxygen for oxidation and decomposition of the organic materials and for the respiration of the microbes and cultured animals. This kind of cycle may improve the nutrient cycle, and it can create a balance between bacteria and micro algae, and maintaining a good water quality environment for the cultured animals.

The Feasibility and Future of the Application of Probiotics in Aquaculture

Based on the previous research results on probiotics we suggest that the use of probiotic bacteria in aquaculture has tremendous scope and the study of the application of probiotics in aquaculture has a glorious future. At present, the probiotics are widely applied in United States of America, Japan, European countries, Indonesia, India and Thailand, with commendable results. The probiotics have become commodities in some countries, for example, the Alken-Murray Corporation and American Standard Products company of United States of America and the company of Japan have their probiotics products. The study may create a new field of industrial products, like the industrial fields of aquaculture product processing and Aquacultural food processing.

China is a large country in aquaculture, but the application and development of the probiotics in Chinese aquaculture is very meager when compared to other countries. In recent years, the diseases of shrimps hindered the development of shrimp culture. The Chinese government has realised the economic value and potential social benefits of the application of probiotics in aquaculture, and has, recently, paid more attention to the study and development of probiotics in aquaculture.

Thus the government has increased the research funds for it. Probiotics principally inhibit the growth and decrease the pathogenicity of the pathogenic bacteria, enhance the nutrition of the aquacultured animals, improve the quality of the aquaculture water and decrease the use of antibiotics and other chemicals; thus decreasing environmental contamination by the residual antibiotics and chemicals. This benefit of probiotics will be long lasting, and the application of probiotics will become a major field in the development of aquaculture in the future.

Application of Probiotics in Aquaculture

During the past 20 years, aquaculture industry has been growing tremendously, especially that of marine fish, shrimps and bivalves. But, as with many other industries, this rapid growth has brought with it the problem of environmental pollution. Contamination of coastal waters due to aquaculture is posing serious concerns among law makers as well as scientists. The coastal environment has been seriously damaged, often resulting in disease outbreaks.

Recently, shrimp culture all over the world has been frequently affected by viral and bacterial diseases inflicting huge loss. In China, the production of shrimps decreased seriously. The production of shrimps was 200,000 tons in 1992, but was only 55,000 tons in 1994. Pathogenic microorganisms implicated in these outbreaks were viruses, bacteria, rickettsia, mycoplasma, algae, fungi and protozoan parasites. For preventing and controlling diseases, a host of antibiotics, pesticides and other chemicals were used possibly creating antibiotic resistant bacteria, persistence of pesticides and other toxic chemicals in aquatic environment and creating human health hazards. Thus, how to improve the ecological environment of aquaculture has become the focus of attention of international aquaculture.

Now, researchers are trying to use probiotic bacteria in aquaculture to improve water quality by balancing bacterial population in water and reducing pathogenic bacterial load. Researchers are increasingly paying more attention to this new approach (ecological aquaculture), and have made considerable headway. This review, on the basis of the new research findings in probiotics applied to aquaculture, analyse and summarize the mechanism of probiotic action in aquaculture.

Use of Immunostimulants in Aquaculture Systems

The contribution of aquaculture to fish production is steadily increasing. The increase would have been much more but for the major constraint of loses in culture production, particularly of shrimp, due to diseases. In India, the loss of shrimp production during 1995-96 due to diseases is estimated to be Rs.600 crores and the loses continue around this level sine then. Loses in production of cultured shrimp have led to the realisation that the goal of aquaculture is not merely to increase production but to make it sustainable. In recent years the application of vaccination in respect to finfish and immuno-stimulants in respect of shrimp/finfish for disease management in aquaculture is being increasingly recognised. Generally, immunostimulants enhance individual components of the non specific immune response but this does not always translate into increased survival. I addition, immunostimulants fed at too high dose or for too long can be immunosuppressive.

The substances of capable of stimulating immune response are the compounds that promote release of from immune effecter cells. Immunostimulants enhance the humoral and cellular response in both specific and non-specific ways. These agents are widely used for impaired immune function and to stabilize the improved immune status. The use of immunostimulants in fish culture or in aquaculture of other species for prevention of diseases is a promising new development.

In general, immunostimulants comprise a group of biological and synthetic compounds that enhance the non-specific defence mechanisms in animals, thereby imparting generalized protection. This protection may be particularly important for fish that are raised in or released into environments where the nature of pathogen is unknown and immunization by specific vaccine may be futile. Several immunostimulants have been evaluated in fin fishes.

Many occasions arise in the course of fish culture that calls for enhancement of immune response. These include strengthening of the normal immune response in order to enhance protection and reduce immunosuppressive conditions. Immunostimulants can be classified into several categories by their origin and mode of action—

1. bacteria and bacterial products,
2. complex carbohydrates,
3. vaccines,
4. immunity enhancing drugs,
5. nutritional factors,
6. animal extracts,
7. cytokines, and
8. Lectins, plant extracts.

Two main procedures for evaluating the efficiency of an immunostimulants are:

1. *In vivo*, eg., protection tests against fish pathogens: and
2. *In vitro,* eg, measurement of the efficiency of cellular and humoral immune mechanism.

Attention is focused on the lymphocyte proliferation test as an adequate method for providing a correct evaluation of the cellular immune condition which can be adopted together with the more commonly used parameters, such as phagocytosis and respiratory burst. It may be mentioned here that the use of immunostimulants in the diets of marine fish and the evaluation of their effect on the immune system of fish has been investigated. As immunostimulants, as such, which can be useful in preventing diseases in land based aquaculture, in pens and hatcheries rarely occurs alone in the natural environment, the subject deserves a discussion here.

Specific and Non-Specific Immunostimulants: Specific immunostimulation is related to the potentiation of the host's immune system towards a unique specific antigen. Vaccination is perhaps the best example of producing specific immunity.

Non specific immunostimulation generally is an attempt to upgrade immunologic capabilities at a time when an animal may be exposed to one or several pathogens and/or be immuno-compromised.

Characteristics of an Ideal Immunostimulants

These can be described as:

1. It should be non-toxic, even at a high dose rate.
2. It should be non-carcinogenic or have long term side effects.
3. At therapeutic levels, it should have a short withdrawal period with low tissue residues.
4. It should stimulate a wide range of non-specific immune responses against bacteria, fungi, virus, protozoa and helminthes.
5. It should be capable of amplifying primary and secondary immune responses to infectious agents.
6. Breakdown products of compound concerned should be either inactive or readily biodegradable in the environment.
7. It should be having defined chemical composition or biological activity.
8. It should be active by oral route and should be stable both in its native state and after incorporation into food and water.
9. It should be compatible with arrange of drugs including antibiotics and anthelmintics, and
10. It should be inexpensive and either tasteless or palatable.

Objectives of Immunostimulation

These are:

1. Promoting a greater and more effective sustained immune response to those infectious agents producing subclinical disease without risks of toxicity, carcinogeni-city or tissue residues.
2. Hastening the maturation of non-specific and specific immunity in young susceptible animals.
3. Enhancing the level of duration of specific immune response, both cell mediated and humoral, following vaccination.

4. Overcoming of immunosuppressive effects of stress and of those infectious agents that damage or interface with the functioning of cells of immune system.
5. Selectively stimulating the relevant components of the immune system or non-specific immune mechanism that preferentially confer protection against micro-organisms. For example via interferon release, especially for those infectious agents for which no vaccines currently exists; and
6. Maintaining immune surveillance at hightened level to ensure early recognition and elimination of neoplastic changes in tissues.

Some Common Immunostimulants

Muramyl Dipeptide: Muramyl dipeptide is a simple glycoprotein, also a purified form of mycobacteria. Its activity includes:

1. Enhancement of antibody activity.
2. Stimulation of polyclonal activation of lymphocytes, and
3. Activation of macrophages.

Levamisole: It is an anthelmintics chemical that has been shown to have some stimulating effect on the immunological reactivity of animals and humans. Activities of this agent are:

1. enhancement of cell mediated cytotoxicity, lymphokine production and suppressor cell function, and
2. Stimulation of pathagocytic activity of macrophages and neutrophils.

Glucans: Glucans are the most popular immunostimulants used in aquaculture. It is derived from yeast cell wall and from certain higher plants. It has excellent immunostimulatory properties and works well when injected or fed to fish.

Yano et al. (1991) showed that ýÿ3-1, 6, branched ýÿ3-1, 3 Glucans were effective in carp. Jenny and Anderson (1993) showed that the use of Glucans increased activity in non-specific defence mechanism and in protection against *Yesinia ruckeri*. Glucan treatment of Atlantic salmon (*salmo salar*) induced protection against *Vibrio salmonicidia*. Several Glucan products such as vitastim, macrogard, are marketed commercially and are used in supplementing fish feeds.

Chitin and Chitosan: Both chitin and chitosan have a major role in aquaculture. They are non-specific immunostimulators which are effective on a short term basis. Anderson & Swicki (1994) administered chitosan to brook trout (*Salvenus fontinalis*) by injection and immersion and found that high levels of protection occurred 1, 2, 3 days afterwards, but protection was greatly reduced by day 14. Injection of chitosan was also more effective than simple immersion.

Actually chitosan is a deacetylation product of chitin. The influence of chitosan on immune response of healthy and cortisol treated rohu was demonstrated. After treatment with chitosan sufficiently higher responses in almost all assays of non-specific immunity was observed in comparison to their healthy control or cortisol treated counterparts respectively without chitosan treatment (Sahoo and Mukherjee, 1999).

In aquaculture, chitosan has been used as an immunostimu-lant for protection against bacterial disease in fish, for controlled release of vaccines, and as a diet supplement (Bullock et al., 2000). Similar dose of chitosan in brook trout has been shown to be immunopotent. It had a higher degree of protection against *A. salmonicida* infection for a short duration. It also gave protection when feeding was done @ 0.5 gm/100gm feed for one week.

Vitamin C and E: both the vitamins are antioxidants. Vitamin C acts as a multiple cell stimulator. Diet supplemented with vitamin C gave protection against *A. salmonicida* in Atlantic salmon.

Vitamin E Stimulates B and T Lymphocytes: The mode of action of vitamin E in enhancing immunity is nuclear but it has been found that supplemental vitamin E may serve as a significant stimulus of immunity in some individuals.

Bacillus Calmette Guarine (BCG): It is a potent cytokine synthesis enhancer. It is actually a live attenuated vaccine strain of Mycobacterium bovis. BCG produces a generalized enhancement of both B cell and T cell mediated responses of phagocytosis and resistance to infection.

Streptococcal Components: These components are potent immunostimulants. Products from *Bordetella pertuosis, Brucella abortus, Bacillus subtilis* and *Klebsiella* pneumoniae all have immunostimulating activities.

Acemannan: It is a complex carbohydrate. It is a potent cytokine synthesis enhancer with anti-tumor and anti-viral activities. It also has the important property of stimulating wound healing.

Lentinan: It is a polysaccharide extracted from a comestible mushroom. *Lentinus elodes* is endowed with anti-tumor activity. Lentinan might act by increasing sensitivity to histamine and serotoxin.

Leaf Extract of Ocimum Sanctum: Effect of leaf extract Ocimum sanctum on:

1. the specific and non-specific immune responses and
2. Disease resistance against *Aeromonas hydrophila* was investigated in *Oreochromis mossambicus.*

It stimulated both antibody response and neutrophil activity. Dietary intake also enhances the antibody response and disease resistance to *Aeromonas.* Possibility of using *O. sanctum* as immunostimulant is used in the maintenance of finfish health in intensive freshwater aquaculture/

C-UP 111: It has immunostimulant activating leucocyte functions. Highest preventive effect has been shown against *Aeromonas* infection in Nile tilapia with improved neutrophil function, in comparison with Glucans and lactoferin. It is the most popular agent in aquaculture system.

Aquatim: This is a kind of immunostimulant developed by the department of Microbiology of the College of Fisheries, Mangalore and now manufactured and marketed by Mangalore Biotech Laboratory, Mangalore under this trade name. It is widely marketed in India, particularly in Karnataka and Goa.

Aerobic Coryneforms: *Propionebacterium acenes* promotes antibody formation when administered as a killed suspension. This bacteria is phagocytosed by macrophage and stimulates cytokine synthesis. This organism has a general immunostimulating action leading to enhanced antibacterial and antitumor activity.

Prophylactic Measures Used in Aquaculture

World aquaculture has grown tremendously during the last years becoming an economically important industry (Subasinghe *et al.*, 2009). Today it is the fastest growing food-producing sector in the world with the greatest potential to meet the growing demand for aquatic food (FAO 2006). Globally, aquaculture is expanding into new directions, intensifying and diversifying. A persistent goal of global aquaculture is to maximize the efficiency of production to optimize profitability.

However, disease is a primary constraint to the growth of many aquaculture species and is now responsible for severely impeding both economic and socio-economic development in many countries of the world. Disease caused by Vibrio spp. and Aeromonas spp. Are commonly episodes of mortality. When faced with disease problem the common response has been to turn to following prophylactic measures.

The following components are undertaken in prophylactic measurre for disease prevention in aquaculture ;

1. Prebiotics
2. Probiotics
3. Vaccine
4. Immunostimulants

Prebiotic

Prebiotics are a non digestible food ingredient that beneficially affects the host by selectively stimulating the growth and/or activity of one or a limited number of bacteria in the colon and thus improves host health. (Gibson and Roberfroid, 1995). In different studies since 1999, many substances have been investigated as prebiotic. Based on the study of Mahious and Ollevier (2005), Fooks *et al.* (1999), and Gibson *et al.* (2004), any foodstuff that reaches the colon, e.g. non-digestible carbohydrates, some peptides and proteins, as well as certain lipids, is a candidate prebiotic. Certain non-digestible carbohydrates seem authentic prebiotics. They include resistant inulin and oligofructose, transgalactooligosaccharides (TOS), lactulose, isomalto oligosaccharides (IMO), lactosucrose, xylo-oligosaccharides (XOS), soyabean oligosaccharides and glucooligosaccharides. From in vivo and in vitro studies, inulin and oligofructose, TOS and lactulose are presently classified as prebiotics. IMO, lactosucrose, XOS, soyabean oligosaccharides and glucooligosaccharides are not considered as functional ingredients since they do not fulfill all criteria for classification as prebiotics. Prebiotics are selectively fermented by Bifidobacteria, Lactobacillus and Bacteroides. Inclusion of prebiotic in the diet has been reported to increase the uptake of glucose (Breves et al., 2001) and bioavailability of trace elements (Bongers and van den Heuvel, 2003).

In later use of prebiotics, they have the binding capacity therefore increasing the absorption of mineral such as calcium, magnesium and iron; these minerals, are not absorbed in the small intestine and so reach the colon, where they are released from the carbohydrate matrix and absorbed.

Prebiotics have been reported to have numerous beneficial effects in fish such as increased disease resistance and improved nutrient availability. The reasons for the different results are not clear yet. It may be due to the different basal diet, inclusion level, type of omnosaccharide, adaptation period, chemical structure (degree of polymerization, linear or branched, type of linkages between monometric sugars), origin of prebiotic, animal characteristics (species, age, and stage of production), duration of use and hygienic conditions of the experiment. If beneficial effects of prebiotics are manifested in fishes, then prebiotics have much potential to increase the efficiency and sustainability of aquacultural production. Therefore, comprehensive research to more fully characterize the intestinal microbiota of prominent fish species and their responses to prebiotics is warranted and survival.

Probitics

The origin of the term probiotic is attributed to Parker (1974). The probiotics were defined as live microbial feed supplements that improve health of man and terrestrial livestock. The gastrointestinal microbiota of fish and shellfish are peculiarly dependent on the external environment, due to the water flow passing through the digestive tract. Most bacterial cells are transient in the gut, with continuous intrusion of microbes coming from water and food. Some commercial products are referred to as probiotics, though they were designed to treat the rearing medium, not to supplement the diet. This extension of the probiotic concept is pertinent when the administered microbes survive in the gastrointestinal tract. Otherwise, more general terms are suggested, like biocontrol when the treatment is antagonistic to pathogens, or bioremediation when water quality is improved. However, the first probiotics tested in fish were commercial preparations devised for land animals.

Though some effects were observed with such preparations, the survival of these bacteria was uncertain in aquatic environment. Most attempts to propose probiotics have been undertaken by isolating and selecting strains from aquatic environment. These microbes were Vibrionaceae, pseudomonads, lactic acid bacteria, Bacillus spp. and yeasts. Three main characteristics have been searched in microbes as candidates to improve the health of their host. (1) The antagonism to pathogens was shown in vitro in most cases. (2) The colonization potential of some candidate probionts was also studied. (3) Challenge tests confirmed that some strains could increase the resistance to disease of their host. Many other beneficial effects may be expected from probiotics, e.g., competition with pathogens for nutrients or for adhesion sites, and stimulation of the immune system. The most promising prospects are sketched out, but considerable efforts of research will be necessary to develop the applications to aquaculture.

What are Aquatic Probiotics?

The concept for aquatic probiotics is a relatively new. When looking at probiotics intended for an aquatic usage it is important to consider certain influencing factors that are fundamentally different from terrestrial based probiotics. Aquatic animals have a much closer relationship with their external environment. There are the big differences between terrestrial and aquatic animals in the level of interaction between the intestinal microbiota and the surrounding environment. On the other hand, potential pathogens are able to maintain themselves in the external environment of the aquatic organisms and proliferate independently of the host (Hansen and Olafsen 1999; Verschuere et al. 2000; Kesarcodi-Watson et al. 2008). The bacterial community composition of the intestinal tract of

aquatic animals is different from that found in terrestrial animals, which the probiotic concept was developed. Man and terrestrial livestock undergo embryonic development within an amnion, whereas the larval forms of most fish and shellfish are released in the external environment at an early ontogenetic stage. These larvae are highly exposed to gastrointestinal microbiota-associated disorders, because they start feeding even though the digestive tract is not yet fully developed (Timmermans 1987), and though the immune system is still incomplete (Vadstein 1997).

Thus, probiotic treatments are particularly desirable during the larval stages (Gatesoupe 1999). The resident microbes benefit from a fairly constant habitat in the GI tract of man and terrestrial livestock, whereas most microbes are transient in aquatic animals (Moriarty 1990). These animals are poikilothermic and their associated microbiota may vary with temperature changes. Salinity changes in the rearing environment will also affect the microbiota and marine finfish are obliged to drink constantly to prevent water loss from the body. A consequence of the specificity of aquatic microbiota is that the most efficient probiotics for aquaculture may be different from those of terrestrial species (Gatesoupe 1999; Kesarcodi-Watson et al. 2008).

Defining probiotics is a challenge – even more so for aquaculture application. Historically, probiotics were defined according to their expected benefits or improvement to the host's intestinal balance. Being concerned with humans and terrestrial animals, probiotics were generally Gram-positive obligate or facultative anaerobes, mostly LAB. Based on the intricate relationship an aquatic organism has with the external environment when compared with that of terrestrial animals, the definition of probiotics for aquatic animals was modified at the end of the last century. Verschuere et al. (2000) defined aquatic probiotics as "Live microorganisms that have a beneficial effect on the host by modifying the microbial community, associated with the host, by ensuring improved use of the feed or enhancing its nutritional value, by enhancing the host response towards disease, or by improving the quality of its ambient environment". This implies a much wider range of microorga-nisms being used as probiotics for aquaculture animals that for terrestrial animals. The above definition is a more holistic and most appropriately defines probiotics for aquaculture.

Probiotics that currently used in aquaculture industry include a wide range of taxa – from Lactobacillus, Bifidobact-erium, Pediococcus, Streptococcus and Carnobacterium spp. to Bacillus, Flavobacterium, Cytophaga, Pseudomonas, Alteromo-nas, Aeromonas, Enterococcus, Nitrosomonas, Nitrobacter, and Vibrio spp., yeast (Saccharomyces, Debaryomyces) and etc. (Irianto and Austin 2002; Burr et al. 2005; Sahu et al. 2008).

Aquatic Probiotics are Mainly of Two Types:

1) Gut probiotics which can be blended with feed and administrated orally to enhance the useful microbial flora of the gut .
2) Water probiotics which can proliferate in water medium and exclude the pathogenic bacteria by consuming all available nutrients. Thus, the pathogenic bacteria are eliminated through starvation (Nageswara and Babu 2006; Sahu et al. 2008).

The first type probiotics are using mainly in finfish aquaculture and the second type in shrimp aquaculture. Commercially available probiotics include pure strains, defined mixture of specific strains, but also consortia of strains and undefined mixtures. Generally, probiotics proposed as biological control agents in fish aquaculture are applied in the feed or as a water additive supplement.

Aquatic Probiotics are Marketed in Two Forms:

1) Dry forms: the dry probiotics that come in packets can be given with feed or applied to water. They have many benefits, such as safety, easy using, longer shelf life and etc. (Decamp and Moriarty 2007);
2) Liquid forms: the hatcheries generally use liquid forms which are live and ready to act. These liquid forms are directly added to hatchery tanks or blended with farm feed. The liquid forms

can be applied any time of the day in indoor hatchery tanks, while it should be applied either in the morning or in the evening in outdoor tanks. Liquid forms give positive results in lesser time when compared to the dry and spore form bacteria, though they are lower in density (Nageswara and Babu 2006).

There are no reports of any harmful effect for probiotics but it is found that the biological oxygen demand level may temporarily be increased on its application; therefore it is advisable to provide subsurface aeration to expedite the establishment of probiotics organisms. A minimum dissolved oxygen level of 3% is recommended during probiotics treatment. The development of suitable probiotics for aquaculture is not a simple task. It requires empirical and fundamental research, full-scale trials as well as the development of appropriate monitoring tools and production under stringent quality control. A performing mixture of probiotic strains can be designed after evaluating the ability of individual strains to grow in low/high salinity under micro-aerophilic or anaerobic conditions, produce various enzymes, and more importantly, produce a range of inhibitory compounds (Decamp 2004).

Vaccine

A vaccine is any biologically based preparation intended to establish or to improve immunity to a particular disease or group of diseases. Vaccines have been used for many years in humans, terrestrial livestock, and companion animals against a variety of diseases.

Vaccines work by exposing the immune system of an animal to an "antigen"—a piece of a pathogen or the entire pathogen—and then allowing time for the immune system to develop a response and a "memory" to accelerate this response in later infections by the targeted disease-causing organism. Vaccines are normally administered to healthy animals prior to a disease outbreak. One analogy used for vaccines is that of an insurance policy (Komar et al 2004). A vaccine, if effective, can help prevent a future disaster from being a major economic drain. But vaccines, like insurance, have a premium, or cost. The producer must weigh the cost in materials and labour against the risk and cost of a disease outbreak to determine whether vaccination is warranted. When actual vaccine effectiveness is also unknown, this makes decision-making even more difficult. Consultation with a fish veterinarian or other fish health specialist will be helpful when examining cost vs. benefit of a particular vaccine.

The ideal vaccine:

a. is safe for the fish, the person(s) vaccinating the fish, and the consumer;
b. protects against a broad strain or pathogen type and gives 100% protection;
c. provides long-lasting protection, at least as long as the production cycle;
d. is easily applied;
e. is effective in a number of fish species;
f. is cost effective; and
g. is readily licensed and registered (Grisez and Tan 2005).

What are the Different Types of Vaccines?

There are many different types of vaccines, and new kinds are continuously under development. Of the types currently in use, the most common are described below. Bacterins are vaccines comprised of killed, formerly pathogenic bacteria. Bacterins stimulate the antibody-related portion of the immune response (i.e., the humoral immune response).

Live, attenuated vaccines are comprised of live micro-organisms (bacteria, viruses) that have been grown in culture and no longer have the properties that cause significant disease. Live attenuated vaccines will stimulate additional parts of the immune system (i.e., a cell-mediated, as well as a humoral [antibody] response). Toxoids are vaccines comprised of toxic compounds that have been inactivated, so they no longer cause disease. An example, used in humans, is the tetanus toxoid vaccine. Subunit vaccines are made from a small portion of a micro-organism (rather than the entire micro-organism) that ideally will stimulate an immune response to the entire organism.

Other types of vaccines in development use even more modern strategies. Examples include recombinant vector vaccines, which combine parts of disease-causing micro-organisms with those of weakened microorganisms, and DNA vaccines. Recombinant vector vaccines allow a weak pathogen to produce antigen. DNA vaccines are composed of a circular portion of genetic material that can, after being incorporated into the animal, produce a particular immune-stimulating portion of a pathogen (i.e., antigen) continuously, thus providing an "internal" source of vaccine material. Other vaccine strategies are also undergoing research and development.

How are Vaccines Given to Fish?

Vaccines are administered to fish in one of three ways: by mouth, by immersion, or by injection. Each has its advantages and disadvantages. The most effective method will depend upon the pathogen and its natural route of infection, the life stage of the fish, production techniques, and other logistical considerations. A specific route of administration or even multiple applications using different methods may be necessary for adequate protection.

Oral Vaccination results in direct delivery of antigen via the digestive system of the fish. It is the easiest method logistically because feeding is a normal, ongoing part of the production schedule. Stress on the fish is minimal, and no major changes in production are required.

Prior to feeding, vaccine is mixed, top-dressed, or bioencapsulated into the feed. To reduce leaching into the water and/or to provide some protection against breakdown of the vaccine by the fish's digestive processes, a coating agent is often used. For small fish (e.g., 1-5 g or less), bioencapsulation may be a preferred method of oral delivery. Live food (rotifers, brine shrimp) is added to a concentrated vaccine solution, and allowed to take up vaccine. This live food is then fed to fry or small fingerlings. Although oral vaccine is the most preferred method, it conveys relatively short immunity (compared to the other methods) such that additional vaccination may be required. In addition, because of the problems involved with getting the vaccine intact through the intestine and adequately stimulating the immune system, there are few commercial oral vaccines available (Komar et al. 2004).

Immersion Vaccination permits immune cells located in the fish's skin and gills to become directly exposed to antigens. These immune cells may then mount a response (e.g., antibody production), thus protecting the fish from future infection. Other types of immune cells in the skin and gills carry antigens internally, where a more systemic response will also develop. Immersion vaccination occurs by dip or by bath. Dips are short, typically 30 seconds, in a high concentration of vaccine. Baths are of longer duration—an hour or more—and in a much lower concentration of vaccine. In practice, dips are logistically more practical for large numbers of small (1- to 5-g) fish. Unfortunately, protection using immersion methods may not last long and a second vaccination may be required (Komar et al. 2004) because smaller, younger fish may have immature immune systems and because this is a more indirect route.

Injection Vaccination allows direct delivery of a small volume of antigen into the muscle (intramuscular (IM) injection) or into the body cavity (intracoelomic [ICe= intraperitoneal or IP] injection), allowing for more direct stimulation of a systemic immune response. Injection vaccines normally include an oil-based or water-based compound, known as an adjuvant, that serves to further stimulate the immune system. Injection is effective for many pathogens that cause systemic disease; and protection—6 months to a year—is much longer than by other methods. Every fish in the population is injected, giving more assurance to the producer. Another advantage is that multiple antigens (for different diseases) can be delivered at the same time. However, vaccination by injection is logistically the most demanding of all three methods. Fish must be anesthetized to minimize stress. Injection requires more time, labour, and skilled personnel. The correct needle size is important. The vaccine may incite a more severe reaction if it is injected into the wrong portion of the fish. And finally, smaller-sized fish (under 10 g) may not respond well to this method (Komar et al 2004).

Have Vaccines been Used in Fish?

Vaccines have been used in food fish, in particular the salmon industry, for approximately 30 years, and are believed to be one of the main reasons that salmon production has been so successful. Vaccination of salmon also dropped the industry's use of antibiotics to a mere fraction of its original use (Sommerset et al 2005). In Norway, for example, in 1987, before widespread use of vaccines, approximately 50,000 kg of antibiotics were used. By 1997, when vaccines had become more routine, antibiotic usage had dropped to less than 1000-2000 kg (Sommerset et al. 2005).

Diagnostic Tools Used in Fish Disease Diagnosis

Diagnosis is a label given for a medical condition or disease identified by its signs, symptoms, and from the results of various diagnostic procedures. The term "diagnostic criteria" designates the combination of signs, symptoms, and test results that allows the clinician to ascertain the diagnosis of the respective disease. It can be definition as "the recognition of a disease or condition by its outward signs and symptoms," or "the analysis of the underlying physiological/biochemical cause(s) of a disease or condition." Disease diagnosis is refers to the various procedures and techniques used to identify the nature of disease and to precisely pinpoint the primary and secondary pathogens involved. Proper diagnosis leads to accurate therapies and avoid indiscriminate use of chemotherapeutics.

Disease diagnosis is an integral part of aquatic health management. Proper diagnosis helps to adopt accurate therapy and avoid indiscriminate use of chemotherapeutics. Information on case history and clinical signs should be carefully used while examining samples for diagnosis.

Importance of Case History

Case history information is very vital for proper disease diagnosis and for taking proper remedial steps. During a disease outbreak, careful scrutiny of the case history information will help to precisely pinpoint the circumstances under which the disease has developed. Some of the case history that are taken in to account are,

- Water quality
- Feeding percentage
- Feed intake
- Fertilization schedule
- Time of last Algal bloom
- Liming details
- Treatment details
- Source of seed, stocking details

Classification of Disease Diagnostic Methods

The disease diagnostic methods can be classified in to following types

- Microscopic diagnosis
- Histological diagnosis
- Microbiological diagnosis
- Immunological diagnosis
- Molecular diagnosis

Microscopic Diagnosis

Without a microscope it is simply impossible to tell the difference between a water quality and a parasite problem. The microscope should be considered the most basic of tools in fish disease diagnosis. Microscopy can be done quickly, but accuracy depends on the experience of the microscopist and quality of equipment. Most specimens are treated with stains that colour pathogens, causing them to stand out from the background, although wet mounts of unstained samples can be used to detect fungi, parasites (including helminth eggs and larvae) and motile organisms. Visibility of fungi can be increased by applying 10% potassium hydroxide (KOH) to dissolve surrounding tissues and nonfungal organisms.

The stain is based on the likely pathogens, but no stain is 100% specific. Most samples are treated with Gram stain and, if mycobacteria are suspected, an acid-fast stain. However, some pathogens are not easily visible using these stains; if these pathogens are suspected, different stains or other identification methods are required because microscopic detection usually requires a microbe concentration of about 1 x 10/ml.

The microscopes that are commonly used in fish disease diagnosis are

1. Light microscope
2. Electron microscope

Light Microscope

The light microscope is the most commonly used microscope in disease diagnosis. In this microscope the scrapings from the fish can be directly visualized or a squash of organ can be prepared for the examination. It can also be used for interpreting histological slides and stained microorganisms.

Histological Diagnosis

Histology is a branch of biology that involves the microscopic examination of thin, stained tissue sections in order to study their structure and function and, in the case of histopathology, to determine changes which may be due to pathogens and disease. Histology and have a central role in disease diagnosis. Stained fish tissue sections are prepared and examined by light microscopy. Tissue changes resulting from infectious or non-infectious disease are identified and described. Immunohist-ochemical methods are also used to detect specific pathogens in tissue sections.

Preparation

Following sampling, fish tissues are placed in an aqueous fixative. This fixative preserves the morphology (structure and chemical constituents) of tissues and cells, so that they are capable of withstanding further preparatory steps without change. It is essential that tissues are fixed within a very short time after death to avoid disintegration of tissues or cells by the action of their own enzymes. Following fixation, tissues are gradually dehydrated to remove any tissue water, using a series of graded alcohols. The tissues are then 'cleared', which involves treatment with a substance that mixes completely with both the dehydrating fluid and the embedding agent. Next the tissues are embedded in molten paraffin wax and cooled to harden the wax so that thin sections can be cut using a microtome and then mounted onto glass microscope slides. The wax is removed from the sections before staining.

Staining

After clearing and rehydration, the tissue sections can be stained using biological stains or dyes. Haematoxylin and eosin (H&E) is the most widely used histological stain because of its ability to reveal a wide range of different tissue components. Gram's stain is a staining method for differentiating microorganisms. The technique is based on the capability of bacteria cell walls to retain the crystal violet dye in the Gram stain during solvent treatment.

The cell walls for Gram positive micro-organisms retain the primary violet as they have a higher peptidoglycan (sugars) and a lower lipid content than Gram negative bacteria. The Periodic Acid-Schiff (PAS) reaction is used to demonstrate certain carbohydrates that are present in some tissues, and provide identification of infecting fungus in fish tissues. The PAS positive sites stain pink/red.

Immunohistochemistry

Immunohistochemical staining methods have been developed for the detection of viruses such as infectious pancreatic necrosis virus (IPNV), infectious salmon anaemia virus (ISAV) and nodavirus in paraffin-embedded tissue sections. Viral antigen is localised by an antibody raised against the virus and subsequent detection steps result in a coloured product that can be visualised by light microscopy

Histological techniques enable the description of tissue pathology and highlight the sequence of cellular changes and their progression caused by infectious and noninfectious diseases. By examining stained sections, viruses, bacteria, fungi and parasites can be identified, and using immunohistochemical techniques, certain infectious agents can be detected in tissue sections. The increased use of image analysis tools by FRS allows qualitative data to be generated to enhance disease diagnosis.

Microbiological Diagnosis

Bacterial diseases can be identified by various microbiology methods. The samples are taken from the fish under aseptic condition and they are first grown in non-selective media, then the various microbial diagnostic methods are used. The different microbiological diagnostic methods are

1. Staining methods
2. Motility test
3. Culturing in selective medium
4. Biochemical test

Staining Method

Various staining methods are used to identify various bacterial pathogens among them gram staining is the most important and most commonly used method.

- *Gram stain:* The Gram stain classifies bacteria according to whether they retain crystal violet stain (gram-positive — blue) or not (gram-negative — red) and highlights cell morphology (eg, bacilli, cocci) and cell arrangement (eg, clumps, chains, diploids). Such characteristics can direct antibiotic therapy pending definitive identification. To do a Gram stain, technicians heat-fix specimen material to a slide and stain it by sequential exposure to Gram's crystal violet, iodine, decolorizer, and counterstain (typically safranin).
- *Acid-fast and moderate (modified) acid-fast stains:* These stains are used to identify acid-fast organisms (Mycobacterium sp) and moderately acid-fast organisms (primarily Nocardia sp). These stains are also useful for staining Rhodococcus and related genera, as well as oocysts of some parasites (eg, Cryptosporidium).
- *Fluorescent stains:* These stains allow detection at lower concentrations (1 10 cells/mL). Examples are acridine orange (bacteria and fungi), auramine-rhodamine and auramine O (mycobacteria), and calcofluor white (fungi, especially dermatophytes). Coupling a fluorescent dye to an antibody directed at a pathogen (direct or indirect immunofluorescence) should theoretically increase sensitivity and specificity. However, these tests are difficult to read and interpret, and few (eg, Pneumocystis and Legionella direct fluorescent antibody tests) are commercially available and commonly used.

- *India ink (colloidal carbon) stain:* This stain is used to detect mainly encapsulated fungi in a cell suspension. The background field, rather than the organism itself, is stained, which makes any capsule around the organism visible as a halo. Leukocytes may appear encapsulated.
- *Wright's stain and Giemsa stain:* These stains are used for detection of parasites in blood, phagocytes and tissue cells, intracellular inclusions formed by viruses and some intracellular bacteria.
- *Trichrome stain (Gomori-Wheatley stain) and iron hematoxylin stain:* These stains are used to detect intestinal protozoa. The Gomori-Wheatley stain is used to detect microsporidia. It may miss helminth eggs and larvae. The iron hematoxylin stain differentially stains cells, cell inclusions, and nuclei. Helminth eggs may stain too dark to permit identification.

Motility Test

Histological techniques enable the description of tissue pathology and highlight the sequence of cellular changes and their progression caused by infectious and noninfectious diseases. By examining stained sections, viruses, bacteria, fungi and parasites can be identified, and using immunohistochemical techniques, certain infectious agents can be detected in tissue sections. The increased use of image analysis tools by FRS allows qualitative data to be generated to enhance disease diagnosis.

There are three methods to see the motility of the bacteria and they are

i. Hanging Drop Method
ii. Semi Solid Agar Method (Craigie)
iii. Three Coverslip Method

Hanging Drop Method

- Place a small drop of liquid bacterial culture in the centre of a coverslip
- Place a small drop of water at each corner of the coverslip
- Invert a slide with a central depression over the coverslip
- The coverslip will stick to the slide and when the slide is inverted the drop of bacterial culture will be suspended in the well
- Examine microscopically (x400) for motile organisms

Note: If well slides are not available, a ring of Vaseline or plasticine may instead be made on an ordinary microscope slide

Positive result: A darting, zigzag, tumbling or other organised movement

Negative result: No movement or Brownian motion only.

Semi Solid Agar Method (Craigie)

- Inoculate the test organism - the central glass tube
- Incubate at the relevant temperature for 18-24 hours
- Subculture from the outer section of the medium

Positive result: Organism can be recovered from the outer section of the medium Negative result: Organism remains in the inner tube

Culturing Bacteria Pathogen in Selective Medium

Certain bacteria have the capacity to grow in selective media. This property of bacteria is used to identify the bacterial pathogen. Only particular kind of bacteria can grow in selective media. Example for selective medium is TCBS for *Vibrios.*

Biochemical Tests

Various biochemical test are used for the identification of bacteria based on the character of a particular bacteria to give its positive and negative result for a particular biochemical test.

1. *The Indole Test:* The test organism is inoculated into tryptone broth, a rich source of the amino acid tryptophan. Indole positive bacteria such as Escherichia coli produce tryptophanase, an enzyme that cleaves tryptophan, producing indole and other products. When Kovac's reagent (p-dimethylaminobenzaldehyde) is added to a broth with indole in it, a dark pink colour develops. The indole test must be read by 48 hours of incubation because the indole can be further degraded if prolonged incubation occurs. The acidic pH produced by Escherichia coli limits its growth.
2. *The Methyl Red and Voges-Proskauer Tests:* The methyl red (MR) and Voges-Proskauer (VP) tests are read from a single inoculated tube of MR-VP broth. After 24-48 hours of incubation the MR-VP broth is split into two tubes. One tube is used for the MR test; the other is used for the VP test. Media contains glucose and peptone. All enterics oxidize glucose for energy; however the end products vary depending on bacterial enzymes.

Both the MR and VP tests are used to determine what end products result when the test organism degrades glucose. E. coli is one of the bacteria that produces acids, causing the pH to drop below 4.4. When the pH indicator methyl red is added to this acidic broth it will be cherry red (a positive MR test). Klebsiella and Enterobacter produce more neutral products from glucose (e.g. ethyl alcohol, acetyl methyl carbinol). In this neutral pH the growth of the bacteria is not inhibited. The bacteria thus begin to attack the peptone in the broth, causing the pH to rise above 6.2. At this pH, methyl red indicator is a yellow colour (a negative MR test). The reagents used for the VP test are Barritt's A (alpha-napthol) and Barritt's B (potassium hydroxide). When these reagents are added to a broth in which acetyl methyl carbinol is present, they turn a pink-burgundy colour (a positive VP test). This colour may take 20 to 30 minutes to develop.

3. *Catalase Test:* Catalase is the enzyme that breaks hydrogen peroxide (H_2O_2) into H_2O and O_2. Hydrogen peroxide is often used as a topical disinfectant in wounds, and the bubbling that is seen is due to the evolution of O2 gas. H_2O_2 is a potent oxidizing agent that can wreak havoc in a cell; because of this, any cell that uses O2 or can live in the presence of O_2 must have a way to get rid of the peroxide. One of those ways is to make catalase.
4. *The Citrate Test:* The citrate test utilises Simmon's citrate media to determine if a bacterium can grow utilising citrate as its sole carbon and energy source. Simmon's media contains bromthymol blue, a pH indicator with a range of 6.0 to 7.6. Bromthymol blue is yellow at acidic pH's (around 6), and gradually changes to blue at more alkaline pH's (around 7.6). Uninoculated Simmon's citrate agar has a pH of 6.9, so it is an intermediate green colour. Growth of bacteria in the media leads to development of a Prussian blue colour (positive citrate). Enterobacter and Klebsiella are citrate positive while E.coli is negative.
5. *Oxidase Test:* The oxidase test identifies organisms that produce the enzyme cytochrome oxidase. Cytochrome oxidase participates in the electron transport chain by transferring electrons from a donor molecule to oxygen. The oxidase reagent contains a chromogenic reducing agent, which is a compound

that changes colour when it becomes oxidized. If the test organism produces cytochrome oxidase, the oxidase reagent will turn blue or purple within 15 seconds.

6. *Nitrate Reduction Test:* Nitrate broth is used to determine the ability of an organism to reduce nitrate (NO_3) to nitrite (NO_2) using the enzyme nitrate reductase. It also tests the ability of organisms to perform nitrification on nitrate and nitrite to produce molecular nitrogen. Nitrate broth contains nutrients and potassium nitrate as a source of nitrate. After incubating the nitrate broth, add a dropperful of sulfanilic acid and a-naphthylamine. If the organism has reduced nitrate to nitrite, the nitrites in the medium will form nitrous acid. When sulfanilic acid is added, it will react with the nitrous acid to produce diazotized sulfanilic acid. This reacts with the a-naphthylamine to form a red-coloured compound. Therefore, if the medium turns red after the addition of the nitrate reagents, it is considered a positive result for nitrate reduction. If the medium does not turn red after the addition of the reagents, it can mean that the organism was unable to reduce the nitrate, or it could mean that the organism was able to denitrify the nitrate or nitrite to produce ammonia or molecular nitrogen. Therefore, another step is needed in the test. If the medium does not turn red after the addition of the nitrate reagents, add a small amount of powdered zinc. Be careful, as powdered zinc is hazardous! If the tube turns red after the addition of the zinc, it means that unreduced nitrate was present. Therefore, a red colour on the second step is a negative result. The addition of the zinc reduced the nitrate to nitrite, and the nitrite in the medium formed nitrous acid, which reacted with sulfanilic acid. The diazotized sulfanilic acid that was thereby produced reacted with the a-naphthylamine to create the red complex. If the medium does not turn red after the addition of the zinc powder, then the result is called a positive complete. If no red colour forms, there was no nitrate to reduce. Since there was no nitrite present in the medium, either, that means that denitrification took place and ammonia or molecular nitrogen were formed.

7. *Urease Test:* Urease broth is a differential medium that tests the ability of an organism to produce an exoenzyme, called urease that hydrolyzes urea to ammonia and carbon dioxide. The broth contains two pH buffers, urea, a very small amount

of nutrients for the bacteria, and the pH indicator phenol red. Phenol red turns yellow in an acidic environment and fuchsia in an alkaline environment. If the urea in the broth is degraded and ammonia is produced, an alkaline environment is created, and the media turns pink. Many enterics can hydrolyze urea; however, only a few can degrade urea rapidly. These are known as "rapid urease-positive" organisms. Members of the genus *Proteus* are included among these organisms. Urea broth is formulated to test for rapid urease-positive organisms. The restrictive amount of nutrients coupled with the use of pH buffers prevent all but rapid urease-positive organisms from producing enough ammonia to turn the phenol red pink.

8. *Phenol Red Broth:* Phenol Red Broth is a general-purpose differential test medium typically used to differentiate gram negative enteric bacteria. It contains peptone, phenol red (a pH indicator), a Durham tube, and one carbohydrate. We use three different kinds of phenol red broths. One contains glucose; one contains lactose, and the last contains sucrose. The objective of the exercise is to determine which organisms can utilise each sugar. Phenol red is a pH indicator which turns yellow below a pH of 6.8 and fuchsia above a pH of 7.4. If the organism is able to utilise the carbohydrate, an acid by-product is created, which turns the media yellow. If the organism is unable to utilise the carbohydrate but does use the peptone, the by-product is ammonia, which raises the pH of the media and turns it fuchsia. When the organism is able to use the carbohydrate, a gas by-product may be produced. If it is, an air bubble will be trapped inside the Durham tube. If the organism is unable to utilise the carbohydrate, gas will not be produced, and no air bubble will be formed.
9. *Casease Test:* Skim milk agar is a differential medium that tests the ability of an organism to produce an exoenzyme, called casease, that hydrolyzes casein. Casein forms an opaque suspension in milk that makes the milk appear white. Casease allows the organisms that produce it to break down casein into smaller polypeptides, peptides, and amino acids that can cross the cell membrane and be utilised by the organism. When casein is broken down into these component molecules, it is no longer white. If an organism can break down casein, a clear halo will appear around the areas where the organism has grown.

10. *Gelatinase Test:* Nutrient gelatin is a differential medium that tests the ability of an organism to produce an exoenzyme, called gelatinase, that hydrolyzes gelatin. Gelatin is commonly known as a component of gelled salads and some desserts, but it's actually a protein derived from connective tissue. When gelatin is at a temperature below 32°C (or within a few degrees thereof), it is a semisolid material. At temperatures above 32°C, it is a viscous liquid. Gelatinase allows the organisms that produce it to break down gelatin into smaller polypeptides, peptides, and amino acids that can cross the cell membrane and be utilised by the organism. When gelatin is broken down, it can no longer solidify. If an organism can break down gelatin, the areas where the organism has grown will remain liquid even if the gelatin is refrigerated.
11. *Lipase Testl:* Tributyrin agar is a differential medium that tests the ability of an organism to produce an exoenzyme, called lipase, that hydrolyzes tributyrin oil. Lipases break down lipids (fats). Tributyrin oil is a type of lipid called a triglyceride. Other lipase tests use different fat sources such as corn oil, olive oil, peanut oil, egg yolk, and soybean oil. Lipase allows the organisms that produce it to break down lipids into smaller fragments. Triglycerides are composed of glycerol and three fatty acids. These get broken apart and may be converted into a variety of end-products that can be used by the cell in energy production or other processes. Tributyrin oil forms an opaque suspension in the agar. When an organism produces lipase and breaks down the tributyrin, a clear halo surrounds the areas where the lipase-producing organism has grown.
12. *Starch Hydrolysis:* Starch agar is a differential medium that tests the ability of an organism to produce certain exoenzymes, including a-amylase and oligo-1,6-glucosidase, that hydrolyze starch. Starch molecules are too large to enter the bacterial cell, so some bacteria secrete exoenzymes to degrade starch into subunits that can then be utilised by the organism. Starch agar is a simple nutritive medium with starch added. Since no colour change occurs in the medium when organisms hydrolyze starch, we add iodine to the plate after incubation. Iodine turns blue, purple, or black (depending on the concentration of iodine) in the presence of starch. A clearing around the bacterial growth indicates that the organism has hydrolyzed starch.

13. *Triple Sugar Iron Agar:* Triple sugar iron agar (TSI) is a differential medium that contains lactose, sucrose, a small amount of glucose (dextrose), ferrous sulfate, and the pH indicator phenol red. It is used to differentiate enterics based on the ability to reduce sulfur and ferment carbohydrates. As with the phenol red fermentation broths, if an organism can ferment any of the three sugars present in the medium, the medium will turn yellow. If an organism can only ferment dextrose, the small amount of dextrose in the medium is used by the organism within the first ten hours of incubation. After that time, the reaction that produced acid reverts in the aerobic areas of the slant, and the medium in those areas turns red, indicating alkaline conditions. The anaerobic areas of the slant, such as the butt, will not revert to an alkaline state, and they will remain yellow. This happens with *Salmonella* and *Shigella*
14. *Decarboxylation Test:* Decarboxylase broth tests for the production of the enzyme decarboxylase, which removes the carboxyl group from an amino acid. Decarboxylase broth contains nutrients, dextrose (a fermentable carbohydrate), pyridoxal (an enzyme cofactor for decarboxylase), and the pH indicators bromcresol purple and cresol red. Bromcresol purple turns purple at an alkaline pH and turns yellow at an acidic pH. We also add a single amino acid to each batch of decarboxylase broth. The three amino acids we test in our decarboxylase media are arginine, lysine, and ornithine. The decarboxylase test is useful for differentiating the *Enterobacteriaceae*. Each decarboxylase enzyme produced by an organism is specific to the amino acid on which it acts. Therefore, we test the ability of organisms to produce arginine decarboxylase, lysine decarboxylase, and ornithine decarboxylase using three different but very similar media. If the organism is unable to ferment dextrose, there will be no colour change in the medium. If an organism is able to ferment the dextrose, acidic byproducts are formed, and the media turns yellow. As the organisms ferment the dextrose, the media initially turns yellow, even when it has been inoculated with a decarboxylase-positive organism. The low pH and the presence of the amino acid will cause the organism to begin decarbox-ylation. If an organism is able to decarboxylate the amino acid present in the medium, alkaline bypro-ducts are then produced. Arginine is hydrolyzed to ornithine and is then decarboxylated. Ornithine

decarboxylation yields putrescine. Lysine decarbo-xylation results in cadaverine. These byproducts are sufficient to raise the pH of the media so that the broth turns purple. If the inoculated medium is yellow, or if there is no colour change, the organism is decarboxylase-negative for that amino acid. If the medium turns purple, the organism is decarbox-ylase-positive for that amino acid.

15. *Coagulase Test:* The coagulase test identifies whether an organism produces the exoenzyme coagulase, which causes the fibrin of blood plasma to clot. Organisms that produce catalase can form protective barriers of fibrin around themselves, making themselves highly resistant to phagocytosis, other immune responses, and some other antimicrobial agents. The coagulase slide test is used to identify the presence of bound coagulase or clumping factor, which is attached to the cell walls of the bacteria. Bound coagulase reacts with the fibrinogen in plasma, causing the fibrin-ogen to precipitate. This causes the cells to aggluti-nate, or clump together, which creates the "lumpy" look of a positive coagulase slide test. You may need to place the slide over a light box to observe the clumping of cells in the plasma. The coagulase tube test has been set up as a demo for you to observe in class. This version of the coagulase test is used to identify the presence of either bound coagulase or free coagulase, which is an extracellular enzyme. Free coagulase reacts with a component of plasma called coagulase-reacting factor. The result is to cause the plasma to coagulate. In the demo, the coagulase plasma has been inoculated with *Staphylococcus aureus* and *Staphylococcus epidermidis* and allowed to incubate at 37°C for 24 hours. *Staphylococcus aureus* produces free coagulase; *Staphylococcus epidermidis* does not. The coagulase test is useful for differentiating potentially pathogenic *Staphylococci* such as *Staphylococcus aureus* from other Gram positive, catalase-positive cocci.

4

Artificial Hybridization

It has been possible to produce fertile hybrids of beluga (Huso huso) and sterlet (Acipenser ruthenus), two species which do not cross in nature due to marked ecological differences. The sturgeons are remarkable for their ability to cross and for their hybrids that are fertile to various degrees. For many years we crossed the sturgeons in most specific combinations in order to produce interspecific and intergeneric hybrids, not only in the first generation, but also in the second and even the third generation (also back-cross and triple-cross). In our work we did not come across any instance in which it was impossible to produce viable offspring.

Without detailed cytogenetic analyses of the causes of fertility and sterility of fish hybrids, I would like to point out that there are more chances of producing completely fertile hybrids if the species crossed have equal chromosome number, which must be homologous and able to conjugate in the course of gametogenesis. When there is unequal chromosome number and violation of the conjugation process, sterility occurs in various degrees.

Until recently the sturgeons and their hybrids were not studied cytogenetically. We have initiated research work on chromosome complexes of parent species and their hybrids, which has already given interesting results. The various species of the sturgeon family such as Huso huso, sterlet (Acipenser ruthenus), starred sturgeon (A. stellatus), ship sturgeon (A. nudiventris) and, probably, some others are characterized by a small number of chromosomes (about 60), whereas the chromosome number of A. guldenstadti is more than twice as much. Therefore, crossing of the first four species in any combination is likely to produce fertile hybrids; crossing each of the four with A. guldenstadti produces hybrids which are either of limited fertility or are completely sterile, the results of abnormal gametogenesis. Hence, before starting work on hybridization, it is necessary to study chromosome complexes of the species to be crossed; this will permit prediction, with a greater degree of probability as to whether or not the hybrids will be fertile or sterile, and will avoid waste of labour and money.

Heredity of Hybrids

Many scientists, I.V. Mitchurin in particular, have shown that development of some characteristics of hybrids, to a great extent, depends on the environment in which the hybrids were bred. This environmental selection of characters is possible due to the great heterozygosity of the hybrid.

Study of morphological characteristics of certain hybrids of the sturgeons, the carps and others (as compared to their parent species) shows that hybrids have an intermediate heredity not only in the first generation, but also in the second, as well as in the generations produced by back-crossing and triple-crossing. In comparison to the first generation, the following generations, often but not always, are characterized by greater variability; however, no typical morphologic segregation followed by return to the initial species has been observed. Hence it was believed that interspecific hybrids possess the so-called permanent intermediate heredity, which does not obey Mendel's law of segregation. But recent experimental research has proved that the type of heredity observed in crossing different species does obey the law of segregation. But the law, revealed by the behaviour of genes, is impossible to observe visually in morphological characteristics. The species crossed are different in many characteristics, and consequently in a great number of genes. In interspecific crossing high polymeric inheritance occur when numerous genes affect the development of some definite characteristics in a similar way. This is why in the second generation a much greater number of combinations is inevitable; and among the mass of intermediate specimens, only a few of them are homozygous, which, in all characteristics show a return to the parental form. However, often such individuals are not found in the experiment.

In intraspecific crossing, (inter-racial, for example) it does not matter which race is represented by the male or female, since the characteristics of the offspring are not affected. That is to say, reciprocal hybrids do not differ from each other.

This is expressed by the formula: A × B = B × A. In interspecific, intergeneric and more remote crossings, reciprocal hybrids may be different, as for example the mule and the hinny which are the reciprocal hybrids between horse and the donkey.

Differences in reciprocal forms of some fish hybrids are revealed in the following: (i) viability, (ii) growth rate, (iii) rates of development of other characteristics of heterosis and (iv) morphological characteristics. Marked difference in viability is observed in reciprocal intergenetic hybrids of crucian carp (Carassius carassius); the offspring of Carassius females crossed with males of other genera of Cyprinidae (for example Carassius @& x Tinca B&) are viable, whereas the reciprocal offspring (of Carassius males) are not viable. Considerable differences in growth rates of fry were found in reciprocal hybrids of the sturgeons. The hybrid possessed some features characteristic of the maternal species, revealing matroclinic inheritance. Morphologic differences between reciprocal hybrids are even more pronouced: in countable characteristics (number of vertebrae, fin rays) these differences are again matrilinear.

Differences of reciprocal forms of hybrids are determined by cytoplasmic differences of the species crossed, resulting in different interaction of the nucleus and the plasma in reciprocal crossing. Genetic predetermination of cytoplasm may lead to incompatibility of the cytoplasm in certain species with chromosomes of some other species. As a consequence, some reciprocal hybrids may be viable, whereas some others may not be. The same is true of the factor determining the manifestation of various degrees of heterosis in reciprocal hybrids. Cytolplasmic heredity is a basis of matroclinic inheritance as well.

The nature of many phenomena in remote hybridization cannot be explained without consideration of gynogenesis as one fundamental problem of the theory. Gynogenesis is a form of sexual reproduction, when in the process of fertilization, the sperm penetrates the ovum and activates it for development without contribution of paternal chromosomes, so that heredity in the offspring is determined by the female pronucleus alone.

As a result, pure matroclinal pseudohybrids are produced, represented as a rule by females only. Some scientists consider gynogenesis to be a special case of parthenogenasis. This is not correct because in parthenogenesis, activation of the ovum is not initiated by sperm but by some other agents.

Gynogenesis occurs in nature and can be demonstrated experimentally. Most interesting instances of natural gynogenesis are Carassius auratus gibelio and the Mexican viviparous fish Mollienesia formosa. They reproduce gynogenetically, spawning with males of some other species. Some experiments on remote hybridization have resulted in so-called hybrid gynogenesis. Our numerous experiments in crossing Carassius carassius females with Leuciscus cephalus males (different subfamilies) have shown that only a small part of the offspring was viable, and these were matrilinear, i.e. gynogenetic due to the exclusion of the male chromosome complex during development. Most of the embryos died at the very beginning of development owing to incompatibility of maternal and paternal hereditary substance.

Of great theoretical and practical interest, is radiation gynogenesis, produced when eggs are fertilized by sperm exposed to X-rays in dosages sufficient to stop its hereditary function, but not its ability to penetrate the egg and activate its development. In fact, in the fertilization of the sterlet (Acipenser ruthenus) eggs by the sperm of the sturgeon exposed to X-rays, or the beluga (Huso huso) eggs by the irradiated sperm of the sterlet, completely matrilinear offspring were produced (no paternal characteristics were observed).

Since gynogenesis produces only females it may be possible to utilise it for control of sexual ratio in fishes, which is of great practical importance. Increase in the number of females of the sturgeon, for example, would be desirable both for commercial fishery and for artificial culture. Experiments in this field should be continued on a larger scale.

Heterosis

Heterosis in fish hybrids is a natural phenomenon observed in various families such as Acipenseridae, Salmonidae, Cyprinidae, Percidae, Centrarchidae and Poecilidae. As a rule heterosis is more prominent in intergeneric hybrids which are not closely related, than in interspecific or intraspecific hybrids. Heterosis is revealed in various degrees. For example, growth rate of the hybrid may exceed that of the parent species (typical heterosis), or sometimes growth rate of the hybrid may exceed that of one of the parent species and be equal to or less than the other parent (incomplete heterosis).

In the second hybrid generation, heterosis usually declines, but in back-crosses of the first generation hybrid with the parent species (in which the characteristics that are of interest to us is more pronounced) the hybrid may exceed both the parent species; that is, the hybrid will show a prominent heterosis. For example, the hybrid of Huso huso x (Huso huso x Acipenser ruthenus) exceeds Huso huso in growth rate. Both in evolution and selection, hybrid forms are of more value if their fertility is not violated.

Heterosis should be used on a greater scale in selectional culture of new, valuable fish and in acclimatization as well as in crossing of hybrids for commercial culture in ponds and other inland reservoirs.

Control of Fish Diseases by Selection

Selection of fish for resistance to diseases may prove, and in some cases has already proved, to be an important means of control of fish diseases. Selection work often necessitates transportation of fish from one area to another and this often leads to introduction of diseases and parasites into pond establishments and commercial waters where previously they were absent. Selective breeding of animals and plants for higher resistance to diseases assumes ever greater importance. Often resistance to a specific disease is dependent on a relatively simple mutation involving one or a few genes. Any gene mutation is accompanied by change in the synthesis of proteins and may cause protein incompatibility of parasite and host (Kirpichnikov et al., 1967).

Selective breeding for higher resistance to disease has been applied in fish culture only in recent years and the data available are therefore limited. Insufficient study of fish diseases, especially the infectious ones, and contradictory observations on immunological reactions in fish have also contributed to this. Until recently some authors considered that fish were not capable of producing antibodies. The studies of recent years show that fish do produce antibodies but not so intensively as the higher vertebrates. Immunobiological reactions of fish are undoubtedly affected to a great extent by water temperature; the higher the water temperature, the more pronounced the reaction. At very low water temperature the processes of antibody formation are almost imperceptible (Krantz et al., 1964; Vladimirov, 1966).

Bactericidal activity of organs and tissues has been discovered in fish as well. It is probably dependent on the presence of such antimicrobial proteins as complement, lysozyme and properdian (Lukyanenko and Mieserova, 1962; Lukyanenko 1965; Bauer et al., 1965). At temperatures lower than 10°C the mechanism of excretion and bactercidal action of organs and tissues predominate over other immune mechanisms (Avetikyan, 1959). Phagocytosis, a mechanism of antibody formation begins to function with increase in temperature (Goncharov, 1963, 1967). Recent data on fish immunity provide selection breeders with a theoretical background for the breeding of disease-resistant fish strains.

Selective Breeding of Disease-Resistant Fish

When a disease occurs in a stock, it does not affect all fish to the same extent; some fish, sometimes large numbers, are not affected. Schäperclaus (1953) cites this example: In the spring of 1950 a small pond in Peitz fish farm was stocked with 665 two-year-old carp which were kept there until early June. In June, an outbreak of infectious dropsy occurred. When the pond was fished out, it was found that 4 percent of the fish were either dead or doomed to die; 54 percent of the fish had external ulcers typical of the disease; and 42 percent showed no symptoms of the disease. Even progeny of a single pair of spawners reared under the same environmental conditions did not all possess the same resistance. Individual specimens may serve as starting points for selection work.

Fish belonging to different strains and forms of any species differ in their resistance to disease. The wild carp (Cyprinus carpio) and especially its Amur sub-species (C. carpio haematopterus) are known for their higher resistance to infectious dropsy than the cultured carp (Karpenko and Sventycki, 1961). Amur wild carp × cultured carp hybrids also possess high resistance. The replacement of cultured carp with hybrid forms in recent years in a number of Ukrainian fish farms have resulted in considerable decrease in the incidence of dropsy in this region.

The same was observed in relation to another widespread infectious disease of carp, the air-bladder disease; it is widespread, not only in the USSR but in the other European countries. Arshaniza (1966) carried out experiments on joint rearing of the cultured carp and the hybrid of the fourth generation of cultured carp × Amur wild carp. The results showed that the incidence of airbladder disease amounted to 78 percent in the former, while it occurred in only about 30 percent of the latter. Symptons of illness were more pronounced in the cultured carp than in the hybrid.

Considering these examples, we may assume that work on development of disease resistant strains of fish may be started not only with the selection of specimens possessing higher individual resistance (fish that remained healthy within a stock affected by the disease), but also with the development of breed groups of hybrid origin, distinguishable by higher resistance to a particular disease. Naturally, such selection work should be carried out at those fish farms which are always under the effect of the disease. Otherwise the increased disease resistance of the selected stock would easily be lost.

As stated earlier, data are scant on the selection of disease resistant stocks of fish, and many of the studies undertaken on the problem have either not been carried through or are still in progress. Only three examples of such studies have been carried to conclusion, two of which concern infectious dropsy. For a long time dropsy was considered a bacterial disease, Aeromonas punctata being its pathogenic agent (Schäperclaus, 1930, 1954). Subsequently, many scientists have questioned the bacterial nature of the disease and have come to the conclusion that dropsy is caused by virus.

Direct evidence of the viral nature of dropsy was obtained several years ago by Yugoslavian scientists (Tomasec et al., 1964). They succeeded in growing the causative agent of dropsy on artificially-cultured tissue of carp kidney, and infecting healthy fish with it. The second International Symposium on Fish Diseases held in Munich in 1965, recognised a virus as the primary pathogenic agent of dropsy, while A. punctata and other saprophytic bacteria complicate the course of the disease.

Breeding of Dropsy-Resistant Strain of Carp

Attempts to create a stock of dropsy-resistant carp were made for the first time by Schäperclaus (1953). Work on selection of dropsy-resistant carp was started by him in 1935. Carp not affected by the disease, in ponds where dropsy outbreak had occurred, were chosen for selective breeding. For two successive years the carp were exposed to large amounts of A. punctata with a view to artificially strengthening their natural immunity.

Fishes having even the slightest symptoms of disease were eliminated. The offspring of both selected and non-selected fish were compared for dropsy resistance; the results showed that of the former only 2 percent contracted dropsy, but of the latter 30 percent did. No dropsy was recorded among the progeny of the selected spawners. From 1939 to 1943 the mortality rate for the ponds (65 ponds with a total area of 180 hectares) where the progeny from selected spawners was reared, averaged 11.5 percent, its range varying between 0 and 37 percent. The average mortality rate for the control ponds (76 ponds with a total area of 474 hectares) was 57 percent, its range being 4 to 96 percent. Decrease in mortality was observed in subsequent years as well. From these observations Schäperclaus concluded that resistance to dropsy is inheritable and it is possible by means of correct selection to obtain stocks possessing relative immunity to this disease. However, it is possible to obtain this effect only where a selected stock is not mixed with any other and where no infection is brought in from outside, either with the water or with the fish delivered to the station. Experience over many years has shown that infection brought into a specially selected stock manifests, as a rule, in an acute form.

But such complete isolation of the selected stock cannot be maintained in the carp farms situated within the region of the natural habitat of Cyprinus carpio. Infection is continually brought in either by the wild fish penetrating into the fish farm, or by the water supply or perhaps by some other unknown agent. It was suggested by Kirpichnikov et al., (1967) that such fish farms should select spawners from breed groups that are most resistant to dropsy and increase their resistance further by means of controlled selection. Such work has been started by Kirpichnikov at the fish farms of the Northern Caucasus. The aim of the first phase of this work was to compare the resistance to dropsy of different breed groups of carp. The following groups of carp have been tested, singly as well as in mixed groups.

i. Ropsha carp (hybrid Amur wild carp × mirror carp, F_4 and F_5),
ii. Ukrainian carp selected by A. Kuzema,
iii. local scaly carp and low-bred mirror carp of unknown origin,
iv. wild carp of the Don river.

The trials for resistance to dropsy were carried out at one of the fish farms. The larvae were reared separately. Fish of different breed groups were combined and reared for three years (1963–1965). The fish were marked by fin clipping; however, under the conditions of a hot summer, the fins grew, causing errors in differentiation of breed groups. The results of the test showed that the various groups showed marked differences in their resistance to dropsy. The Ropsha carp, local scaly carp and Don River wild carp proved to be the most resistant to this disease. The enhanced resistance of the above groups is apparently connected to their wild carp origin. Besides high resistance to dropsy, the Ropsha carp was characterized by generally higher viability, and this carp was therefore selected for breeding at the fish farms of the Northern Caucasus, where outbreaks of dropsy often occur. In the spring of 1966, three-year-olds of all groups had attained sexual maturity and served as brood stock of the second generation, which also was subjected to intensive selection for resistance to dropsy. Though this work is far from completion it promises great economic possibilities, especially in the Northern Caucasus, where dropsy is a major hazard in fish culture.

Breeding of Disease-Resistant Trout

In the early fifties, work was started on the creation of trout stocks (the brown trout, Salmo trutta m. fario and the brook trout, Salvelinus fontinalis) resistant to two widespread bacterial diseases, furunculosis and ulcer disease (Wolf, 1954). The initial phase of the work was to test the susceptibility of different fish populations to these diseases. For this, ponds at one trout farm were stocked with fish brought from different bodies of water. Fish affected by the diseases were released into these ponds and the dead fish were fed to the experimental ones. The diseases broke out in all experimental ponds, causing mortality. In the first year, the mortality rate of the brown trout in the experimental ponds varied from 35.7 to 65.8 percent; that of the brook trout fluctuating between 5.3 and 99.6 percent. Thus the two species were seen to differ in their resistance to the diseases. Populations with high resistance were chosen for further selection work. Results obtained in subsequent experiments are not available in the literature accessible to us.

Further work on selection of trout for resistance to furunculosis has been reported by Elinger (1964). Diverse geographical strains of the brown trout and the brook trout were intensively exposed to Aeromonas salmonicida, and the less resistant individuals were removed by mass selection. Survivors were selected as brood fish for the next generation, which was similarly exposed to the disease. Hybridization through crossing more promising strains in some instances improved resistance. Resistant strains were compared with regular hatchery stock under actual hatchery conditions. Results obtained thus far indicate that resistance to furunculosis is being achieved. The work is in progress.

Effect of Selection Work on the Distribution of Parasites and Diseases of Fishes

For carrying out selective breeding of fish it is necessary to obtain fish belonging to different populations, forms, subspecies, or species from different regions for rearing in experimental fish farms. It is apparent that pathogenic agents of different diseases might be brought in with these fish. The danger of bringing in infection is negligible when eggs are chosen as the initial material for selection. There is only one instance on record where infection was introduced into a hatchery with eggs; the pathogenic agent of whirling disease, one of the most troublesome trout diseases, Myxosoma cerebralis, was introduced with trout eggs.

It was supposed that the spores of the organism could stick to the trout eggs and be introduced in this way into new fish farms and bodies of water. However, there is no corroboration of the possibility of such introduction. Efforts in obtaining carp eggs under hatchery conditions, with a view to propagate them artificially, have shown that healthy progeny may be obtained from the eggs of diseased fish. It has been demonstrated that such widespread disease of carp as infectious dropsy and air-bladder disease are not transmitted germinatively. However, there exists the possibility of introduction of parasites infesting the eggs into new bodies of water and fish farms. Fortunately, such parasites are few. Only one species of Coelenterata, infesting acipenserid eggs (Polypodium hydriforme) has been recorded in the U.S.S.R. The danger of bringing it along with eggs is not excluded, and every precaution should be taken while transporting eggs for fish-cultural purposes.

The danger of possible introduction of pathogens while transferring fry or older fish is more real. Soviet ichthyo-pathologists have collected a good deal of data on this. The following is one example. From 1958 to 1962, large numbers of fry of the grass carp Ctenopharyngodon idella, caught in the rivers of China, were transferred to the U.S.S.R. In spite of all the precautions taken, more than 20 species of parasites were introduced with these fish. Many of these parasites cause severe epidemics, not only among phytophagous fish, but among other cultured and commercial fish as well. Thus the cestode Bothriocephalus gowkongensis came to be introduced into the fish-farms of the European part of the U.S.S.R. and Central Asia; then it infected the carp and many other fishes, causing high mortality. It has since penetrated into many bodies of water (Musselius, 1967), and has been introduced into Ceylon (Fernando and Furtado, 1962) and into Romania and other countries of eastern Europe (Radulescu and Georgescu, 1964).

There are many examples of accidental introduction of pathogenic agents of fish diseases along with fish brought in for selection work. The selection of the cold-resistant strain of carp is based on the crossing of Amur wild carp and the cultured carp (Cyprinus carpio × C. carpio hematopterus). In the course of this work a large number of parasites of Far-Eastern origin came to be introduced into the various fish farms.

Back in 1936 the first lot of Amur wild carp was brought in, resulting in the introduction of two monogenetic trematodes which parasitize the gills, and a cestode. Dactylogyrus extensus and Khawia sinensis cause severe diseases, and at present it is difficult to find a fish farm where these do not occur.

In 1949 some Amur wild carp were introduced into fish farms in the northwestern regions of the U.S.S.R. and Byelorussia. Some of these fish were infected with the parasitic protozoan, Ichthyophthirius multifiliis, which was unknown in the carp ponds of the U.S.S.R. Long-distance transportation which lasted about 20 days promoted the development of further infection and when transferred to fish farms, the wild carp passed on the infection to the local spawners. In the spring of 1950 a heavy infection caused high mortality among the spawners. At the Velikolusky fish farm 100 percent of the spawners were lost, and the mortality in the Byelorussian farms was not less than 50 percent. The disease was introduced with the stocking material into the other farms of Byelorussian and the northwestern regions of the U.S.S.R., and by the mid-fifties it had spread to the fish farms of the Ukraine, Central Asia and Kazakhstan.

It was proved experimentally that fish exposed to infection for the first time turned out to be carriers of the largest numbers of Ichthyophthirius. The intensity of disease is 20 times less in fish exposed to the infection repeatedly and further exposures helped to develop immunity to the disease (Bauer 1955, 1958). Toward the end of the fifties a new lot of the Amur wild carp was introduced into three fish farms of Latvia for selection work. In a year a severe outbreak of dropsy occurred. It was also found that a large nematode, previously unknown in the U.S.S.R., was brought in with that lot of fish. Vismanis (1962 and 1967) described the nematode as a new species, Philometra lusiana. Large females of this nematode, parasitizing under carp scales, cause the formation of ulcers and consequent deterioration of the market quality of the fish.

In 1966 a new lot of Amur wild carp was transferred to Ropsha, an experimental fish farm of the State Research Institute of Lake and River Fisheries; every precaution had been taken. In the autumn of 1967, myxosporidians of the genus Sphaerospora were found on the gills of carp in a number of ponds; this parasite had never been found in Ropsha ponds before.

It is on record that fish in a number of fish farms were infected by air-bladder disease, introduced there with the carp broodstock brought in for selection purposes. Thus any transfer of fish for the purpose of selection is always accompanied by the danger of introduction of pathogens, even though every precaution is taken.

Only introduction of fish at the egg or at the larval stage (during the first few days after hatching) eliminates this danger. This is possible with adequately worked-out methods of artificial spawning. For salmonids and whitefish such methods have been developed since the close of the last century and, as a rule, these species are transplanted at the egg stage. Only this may explain the fact that not a single pathogen has been introduced with rainbow trout exported from America to Europe and other continents. At present there are well-developed methods for obtaining fertilized eggs of Acipenseridae and phytophagous Cyprinidae. The eggs of the latter are pelagic. Until recently no effective method was known for taking and incubating adhesive eggs as those of carp. Removal of the adhesive material of carp eggs by any of a number of recently developed methods would enable artificial reproduction of this species, as well as long-distance transportation of the eggs and larvae. In recent years the possibility of obtaining healthy progeny from carp spawners affected by infectious diseases and parasites has been examined. Thus in 1965 (in Byelorussia) and 1967 (in Yazhelbyzy) eggs were obtained from carp spawners affected by air-bladder disease. These eggs were incubated in sterile water, and the larvae were released into ponds where the disease had not been recorded, or to carefully disinfected ponds supplied with water from a non-infected body of water. When mass examination of fingerlings was carried out, not a single individual was found affected by the disease.

Artificial reproduction of wild carp has been carried out at the Tsymlansk hatchery-nursery for a number of years and in 1967 about 70 million larvae were obtained. Observations have shown that if larvae are released into carefully disinfected ponds, they will be practically free of parasites when fished out in autumn. Transporting of fish at the egg and larval stage delays the process of obtaining progeny from the transferred fish by several years. But the delay is worthwhile for it excludes the danger of introduction of infectious diseases with fish brought in for selection purposes.

Rine Biotechnology : Bioactive Natural Products and Their Applications

Marine (or blue) biotechnology encompasses the applications of biotechnology tools on marine resources. Marine biotechnology encompasses those efforts that involve the marine resources of the world, either as the source or target of biotechnology applications. Biotechnology is the application of science and technology to living organisms, as well as parts, products and models thereof, to alter living or non-living materials for the production of knowledge, goods and services'. In the case of marine biotechnology, the living organisms derive from marine sources.

Biotechnology is defined as the industrial use of living organisms or biological techniques developed through basic research; marine biotechnology is an emerging discipline based on the use of marine natural resources. The oceans encompass about 71% of the surface of our planet, but over 99% of the biosphere (since organisms are found throughout the water column), and they represent the greatest extremes of temperature, light, and pressure encountered by life. Adaptation to these harsh environments has led to a rich marine bio- and genetic-diversity with potential biotechnological applications related to drug discovery, environmental remediation, increasing seafood supply and safety, and developing new resources and industrial processes.

Drug discovery represents one of the most promising and highly visible outcomes of marine biotechnology research. Biochemicals produced by marine invertebrates, algae and bacteria, are very different than those from related terrestrial organisms and thus offer great potential as new classes of medicines. To date, examples of marine-derived drugs include an antibiotic from a fungi, two closely related compounds from a sponge that treat cancer and the herpes virus, and a neurotoxin from a snail that has painkiller properties making it 10,000 times more potent than morphine without the side effects. However, there are several more marine-derived compounds currently in clinical trials and it is likely that many more will advance to the clinic as more scientists look to the sea for these biotechnological uses.

In addition to new medicines, other uses for marine-derived compounds include: cosmetics (algae, crustacean and sea fan compounds), nutritional supplements (algae and fish compounds), artificial bone (corals), and industrial applications (fluorescent compounds from jellyfish, novel glues from mussels, and heat resistant enzymes from deep-sea bacteria).

Marine biotechnology may include techniques such as bioprocessing, bioharvesting, bioprospecting, bioremediation, using bioreactors etc (so called process biotechnology techniques);

aquaculture/fisheries; gene, protein, or other molecule based techniques; while applications may include: health, food, cosmetics, aquaculture & agriculture, fisheries, manufacturing, environmental remediation, biofilms and corrosion, biomaterials, research tools etc. Therefore, marine biotechnology has a horizontal scope encompassing very different applications, for all of which the marine environment is providing the resources. This means everything from deriving a new cancer treatment from a deep-sea sponge to developing an innovative buoy system for monitoring ocean pollution. Like the broader field of biotechnology, marine biotechnology (marine biotech or MBT) can take very traditional forms, such as localized seaweed farming, and high-tech forms. The full range of genomic tools, for instance, is now being applied to such goals as determining precisely how a promising compound derived from a marine organism kills cancer cells. Commercial applications of marine biotechnology includes bio-prospecting, improving the production of marine organisms, production of novel products, particularly food and feed products and diagnostics and biosensors. Bioprospecting includes biotechnology use in whole drug/molecule development process, i.e. screening, identification, efficacy testing, safety testing, large scale commercial production. Bioactive natural products are secondary metabolites produced by organisms that live in the sea. These products have received increasing attention from chemists and pharmacologists during the last two decades. These products have been exploited for variety of purposes including use as food, fragrances, pigments, insecticides and medicines. Through improved biological screening method the role of these products in drug discovery has been greatly enhanced in the last few years. These products show an interesting array of diverse and novel chemical structures with potent biological activities.

Research into pharmacological properties of marine natural products has led to the discovery of many compounds considered worthy of clinical applications. There are great potential in bioprospecting from the sea and marine natural products research has just started to bloom. Today, marine sources have the highest probability of yielding natural products with unprecedented carbon skeletons and interesting biological activities. Many more prospects regarding new habitats for example deep ocean samples and symbiotic systems are still wide open for research.

Bioprospecting

It is high-throughput screening for novel compounds, especially drugs (other uses includes in foodstuffs, nutraceu-ticals, adhesives, paints, cosmetics, environmental remediation, research etc). It includes biotechnology use in whole drug/molecule development process, i.e. screening, identification, efficacy testing, safety testing, large scale commercial production. Traditionally only 1 out of 10,000-20,000 molecules extracted from terrestrial micro-organisms, plants or animals finally reached the market, which may take 10-15 years and cost up to $800m12,13. This has resulted in large pharmaceutical groups abandoning their search for new drugs derived from natural substances. Evidence of complexity in this process is the relatively empty pipeline of pharmaceutical companies.

However, marine biota present a better opportunity for encountering successful candidates in view of the large biodiversity, lack of current knowledge and extreme environments. Ara-C and Ara-A (estimated annual worth $50-100 million) drugs were developed from sponges in the early 50s and proved to be commercial success stories. Anti-cancer agents from marine organisms have an estimated value of $1 billion a year. Another good example includes Vent-DNA polymerase which is used as a basic constituent in Polymerase Chain Reaction (PCR). Bioprospecting is applied to all animal and plant phyla living in shallow as well as in deep seabed ecosystems. The advantages of the first include less technical complexity and better economic viability whereas for the latter a larger and perhaps more interesting natural resource base. In general, it is thought that the pace of discovery of new species and products that are potentially useful to pharmacology is higher for marine and microbial than for terrestrial organisms.

Marine Bioprospecting: Mining the Untapped Potential of Living Marine Resources

Compared to the high degree of representation of terrestrial-derived bioproducts, the number of marine natural products that have found their way into hospitals, clinics and pharmacies is thus far small. This has more to do with the relative infancy of the field (compared to terrestrial bioprosp-ecting) than any lack of potential for discovery. In fact, the natural products isolated from marine sources tend to be more highly bioactive than terrestrial counterparts. This is in part because they have to retain their potency despite dilution in surrounding seawater to be effective in the "chemical warfare" that allows organisms such as sponges to ward off would-be predators and animals that might attempt to grow over and smother them. Despite lesser attention paid to marine natural products historically, there are notable marine-derived bio products that are commercially available, including:

- The family of antiviral drug, including Aids treatment AZT, based on a group of compounds (arabinosides) extracted from the sponge *Tethya crypta* more than 40 years ago. This success story from marine biomedicine's first wave represents an annual market of more than $50 million.
- The anti-inflamatory and analgesic pseudopterosins isolated from a Bahamian soft coral (*Pseudoterigorgia elisabethae*). This led to the development of bioproducts now used in Estee Lauder skin care and cosmetics lines and currently worth $3-4 million a year.
- Ziconitide, known by its trade name Prialtýÿ is a synthetic form of a compound extracted from the venom of predatory tropical cone snails (Conus spp.). Prialtýÿ was approved by the FDA in late 2004 as a treatment for severe cases of chronic pain in patients with conditions such as cancer and AIDS. Current clinical results suggest that Prialt is a powerful, non-addictive alternative to drugs such as morphine

Easily overlooked in the success stories is the technical difficulty of collecting and processing diverse marine organisms, screening them for bioactivity, isolating and identifying natural products, securing a sustainable source of the product, and various other hurdles on the path toward developing a potentially profitable bioproduct.

There is a substantial cost associated with each stage of the endeavour. Furthermore, even when no expense is spared, and even if the best and brightest researchers are on the job, there is never a guarantee that the effort will pan out, but the vast potential payoffs, not only financially through potential commercial sales but also at the human level in lives potentially saved or improved, keeps the field moving forward. Marine bioprospecting is often used to describe the process of collecting marine biota for natural product screening. Of course, biomedical MBT investigations are usually carried out by large research institutions and consortiums and not private individuals, so when a drug lead doesn't 'pan out,' personal financial ruin is not typically the result. Still, the fact that only a handful of the 10,000 or so novel chemical compounds thus far isolated from marine organisms have made their way into the drug discovery pipeline says something about the chances of any one compound ever making its way to the drug market.

Moreover, of those compounds that do enter the clinical development pipeline, a large number eventually are discarded to the biomedical boneyard. As an indication of the investment involved in marine bioprospecting, consider that the UN Atlas of the Oceans reports Japan on its own spends nearly $1 billion on such efforts. Of this sizeable figure, 80% is funded through the private sector.

Specimen Acquisition

A 'typical' marine biotechnology natural products research group will have research objectives centred on any and/or all of the following:

- Collection of marine organisms with promising or as yet unstudied bioactivity
- Detection of bioactivity
- Isolation and identification of bioactive compounds
- Evaluation of the pharmaceutical or other potential of novel compounds

The collection of marine organisms allows assessment of their biomedical or biotechnology potential. The collecting (prospecting) phase of the research is, by definition, a field-based endeavour. The follow-on biotechnology assessment phase is then carried out back in the laboratory in the days, weeks, months, and even years following initial collection.

Collection techniques include:

- Traditional snorkel and SCUBA
- Ship-based collection via bottles, nets, trawls, and benthic grabs, sleds, and dredges
- Collection via ROVs (Remotely Opertated Vehicles) and AUVs (Autonomous Underwhater Vehicles)
- Submersible-based collection

CadalminTM Green Mussel Extract (GMe)

CMFRI was developed new product CadalminTM Green Mussel Extract (GMe) to combat arthritis. CadalminTM. GMe contains 100% natural marine bioactive anti-inflammatory ingredients extracted from green mussel Perna viridis. The product is effective to combat chronic joint pain, arthritis/ inflammatory diseases, and improves cardiovascular functioning. It is an effective green alternative to synthetic non steroidal anti-inflammatory drugs (viz., aspirin containing drugs having undesirable side effects).

The active principle in CadalminTM GMe effectively inhibits inflammatory cyclooxygenase-II and lipoxygenase-V, and its activity was found to be comparable to the drugs available in the market. Consuming CadalminTM GMe will avoid unfortunate side effect of these synthetic non steroidal anti-inflammatory drugs.

This product is a blend of nutraceutical and nutritional elements. CadalminTM GMe is designed to find a unique way to prevent the degradation by air, moisture, heat and light and to maximize the activity.

The product is free from deleterious trans fatty acids, free radicals/ free radical adducts, and low molecular weight carbonyl compounds.

5

Shrimp Culture Practices and Production

Share of Shrimp Aquaculture in the Fisheries Sector

The fishery sector is an important part of Sri Lanka's economy and contributed around 2.2% of GDP in 1993. Fish is the most important source of animal protein consumed in the country, about 60% of the population depending solely on fish for their protein requirement, and accounting for nearly 65% of the total animal protein consumed. The steady growth in local production and increased fish imports led to a rapid increase in *per caput* supplies from 10.7 kg in 1977 to 16.9 in 1989. The bulk of this production came from the coastal marine fishery where in some areas there has been over-exploitation. Inland fish production is heavily dependent on capture fisheries in the reservoirs and tanks.

The marine offshore/deepsea fishery is at present more or less confined to the western and southern coasts and has shown a steady growth from an estimated 3 000 mt in 1978 to 33 000 mt in 1993. The sector employs around 150 000 in fishing, supporting a fishing population of about 600 000. In addition, around 13 500 fishermen were involved in inland fishing and another 50 000 are estimated to be engaged in supporting activities, including fish processing and marketing. Employment in fishing represents about 1.75% of total employment, and about 4% of agricultural employment. When ancillary activities are also considered, the respective shares amount to about 2.5 and 5%. The contribution of shrimp aquaculture production to the total fish production increased from 0.06% (100 mt) to 0.7% (1200 mt) between 1985 and 1993. It has created more than 4 000 jobs, or around 2.0% of fisheries sector employment.

Growth of the Shrimp Culture Industry

Interest in brackishwater shrimp culture in Sri Lanka started in 1977, but the commercial production first entered the market in 1984. In 1984, only one multinational company was producing brackishwater shrimp for the export market. The number of operational establishments had risen to 52 in 1989 and by the end of 1994, the number of authorized farms had increased to 250. The total number of farms including unauthorized farms far exceeds this number.

The water area under culture (excluding bunds) increased from 3 to 12 ha from 1984 to 1986). Due to the expansion of semi-intensive culture practice, the water area under culture increased rapidly in 1997 to 125 ha.

By 1989 the area under culture had increased to 243 ha; since then, no proper records have been maintained. The area allocated, however, for authorized shrimp farms amounted to 1 400 ha at the end of 1994. Once again, the total area developed, including unauthorized farms, far exceeds this total and is distributed along the northwestern coastal belt. The estimated employment generated through shrimp farming was over 4 000 at the end of 1994.

Shrimp Culture Practices

Thirty-one species of shrimp have been recorded from Sri Lankan waters. Of these, only the penacid shrimps have commercial value (de Bruin, 1970). The operational shrimp farms in Sri Lanka practice monostock monoculture of *Penaeus monodon* (black tiger shrimp), a species which has been selected for culture due to its large size, fast growth, high price, high market potential and the availability of technology (Siriwardena, 1990 - unpubl. doc.).

The shrimp farming industry in Sri Lanka is dependent entirely on hatchery-bred-post-larvae. The present requirement of hatchery-bred post-larvac is around 450 million per year, while the present hatchery production is very inadequate, being around 200 million post-larvae per year (Siriwardena, 1990 - unpubl. doc.).

Culture practice in the mid-1980s was largely intensive; however, this trend changed after the disease outbreak in 1988/89. The farmers who were committed to intensive culture used reduced stocking densities and two crops per year, thereby allowing time for treatment. In late 1989, only seven of 52 establishments were practicing intensive culture (Siriwardena, 1990 - unpubl. doc.). The production rates for intensive and semi-intensive culture ranged between 0–15 and 6–8 mt/ha/yr, respectively. The present trend is to practice semi-intensive culture employing a stocking rate of 15–20 post-larvae/m^2 with aeration and artificial feeding.

Of the three sizes of shrimp farm, the small sector comprises the largest number. These farms are viable, and their numbers are expected to increase because of government policy not to permit the existing large farms to become larger and not to approve any more large-scale shrimp farms (Joseph, 1993).

Shrimp Culture Production

Government Plans

Shrimp and finfish culture is being accorded high priority in the National Fisheries Development Plan for 1995 to 2000. Production of cultured shrimp is projected to increase to 5 500 mt, from the present production of 1 200 mt (wet weight).

Shrimp Culture Production and Exports

Shrimp comprises the largest quantity of any export product in the fisheries sector and earns the highest foreign exchange. The percentage contribution of shrimp to the total value of all aquatic products exported varied between 48.5% and 70.3% between 1985 to 1992. The decline in production in relation to the water area under culture can be attributed to the progressive deterioration of the pond environment due to intensive culture and conversion to semi-intensive culture. The contribution of cultured shrimp to total shrimp exports in terms of quantity and value increased from 1985 to 1993, as a capture fishery did not take place in the north and east due to civil disturbances.

Health and Environmental Considerations

Health and environmental cocerns have become major issues in many countries where aquaculture has developed to a commercial scale. These concerns include disease outbreaks, effluent discharge, use of antibiotics, habitat change and damage, and the adverse effects of pollutants on the aquatic environment.

Problems Related to Health Management

The first disease outbreak which significantly affected production occurred during 1988 to 1989. The following disease signs were observed (Jayasinghe, 1995):

- Microfouling on shells
- Reduced frequency of molting
- Reduced feeding
- Black gills
- Soft shell conditions
- Tail rot
- Black spot
- Red/brown deposits on the belly and red colouration.

A drop in production from 5.3 to 1.9 mt/ha/cycle causing an average reduction of 64% in production and an estimated total loss of 186.62 Rs. million (US$ 4.44 million) was observed due to this disease outbreak (Jayasinghe, 1995). No proper monitoring program for health management and disease has been undertaken since the inception of this industry. Hence, estimation of losses due to disease in terms of reductions in quantity and value is difficult. Brown and black gill conditions and MBV virus were reported in grow-out facilities. Iron hydroxide deposits, hypertrophic changes and hemolytic infiltrations were found among gill lamellae, plus infestations by the ectocommensal protozoan *Zoothamnium* (Jayasinghe, 1995; Wijegoonawardena, unpubl. data).

Total bacterial counts of 2.4×10^4 to 9.0×10^5 and 5×10^2 to 8.8×10^3 were reported for cultured shrimp and grow-out pond water, respectively (Fonseka and Hettiarachchi, 1991). Of this, 19.4% and 5% were *Vibrio* and *Pseudomonas*, respectively. These bacterial counts, however, were made for post-harvest technological purposes and it is difficult to make any inferences on the influence of these bacterial levels during culture. Luminescent vibriosis has been recently recognised as the most serious disease affecting hatchery production of *P. monodon*. Luminescent strains of *Vibrio* species have been implicated in outbreaks of shrimp hatchery bacterial diseases mainly in the northwestern coastal areas of Sri Lanka.

The high levels of luminescent vibrios in the sea water of this area may be due to high organic matter, as according to Shilo and Yetinson (1979) these bacteria grow very well in eutrophic sea water. According to hatchery operators of this coastal area, water quality parameters such as temperature, salinity, pH, ammonia, nitrate and phosphate have no effect on the occurrence of luminescent vibrios.

In other countries, two species, *Vibrios harveyi* and *V. splendidus*, have been identified, with *V. harveyi* being more dominant. Significant larval mortalities associated with luminescent vibriosis caused by *V. harveyi* and *V. splendidus* were reported from hatcheries raising *P. monodon* and *P. merguiensis* in Indonesia (Sunaryanto and Mariam, 1986) and in the Philippines (Baticados et al., 1990). The strain(s) of *Vibrio* causing mortalities in hatcheries in Sri Lanka has yet to be determined. In Sri Lanka, vibriosis has been noted to cause up to 100% mortality of *P. monodon* from zoea and mysis to post-larval stages. This has aggravated the problem of production shortage in shrimp post-larvac, leading to the recent importation of *P. monodon* post-larvae into the country. Such importations without proper quarantine measures may cause serious disease outbreaks in the shrimp farming industry.

While most disease control programs emphasize the pathogen (*e.g.*, microbial diagnosis, vaccine development and chemotherapeutic treatment), one must not forget that poor water quality and inadequate nutrition are often basic determinants of disease outbreaks (Sindermann and Lightner, 1988). Questions have been raised as to whether the inter-connected lagoon complex between Chilaw and Puttalam can continue to sustain more than 1 200 ha of shrimp ponds. The canal that serves as the main source of brackish water for the farms is apparently polluted and the risk of disease outbreaks is increasing There is little information on the amounts of different chemotherapeutants used in the shrimp farming industry, but the increase in their use is causing concern, both from human health and the environmental aspects. The commonly use antibiotics are chloramphenicol, oxytetracycline, terramycin and nitrofuran.

Problems Related to the Environment

In brief, the problems related to the environment in the shrimp farming industry in Sri Lanka arose mainly from over-emphasis on high production, economic viability and foreign income generation without full consideration of the environmental impacts caused by over-crowding of farms. At the inception, there was no proper zoning plan to facilitate development of an environmentally sound industry. Lack of a proper zoning plan for the northwestern and western provincial coastal belts led to many social problems and destruction of ecologically sensitive areas, such as mangroves and mud flats. Moreover, this destruction has not yet been quantified in terms of ecological importance and value.

When approving a particular site, neither the industry as a whole nor the existing farms in the vicinity were taken into consideration.

This led to the establishment of farms in a haphazard manner so that the outlets and inlets of individual farms were located in close proximity, inevitably allowing one to obtain the discharged water from the other and leading to health management problems.

The main water sources serving the present shrimp farming industry are Negombo Lagoon, Chilaw Lagoon, Puttalam Lagoon and the canal connecting these three lagoons, the Dutch Canal. According to the renewal rates and quality of water measured in early 1990, the canal, extending from Chilaw to Mundal has the capacity to supply water to a culture area of 242 ha (Siriwardena and Dayaratne, 1990 - unpubl. doc.). Siriwardena and Dayaratne (1990 - unpubl. doc.), recommended improvements to the Dutch Canal in order to enhance the water renewal rate if the shrimp farming industry was to be expanded in the northwestern coastal belt. However, since 1990, the number of shrimp farms has increased rapidly in this area, exceeding the culture area of 242 ha without any improvements in water renewal rates of the Dutch Canal. Many water quality parameters of the Dutch Canal changed due to effluent discharge from shrimp farms. Increased suspended solids may cause a further reduction in the volume and renewal rates of water in the Dutch Canal. If the dissolved oxygen depletes, the increased levels of total sulfides may cause environmental stress leading to fish kills. Such fish kills were reported in the Dutch Canal in early 1995. Moreover, reduced growth rates, lower production and disease problems in shrimp may be attributed to the stress caused by the polluted water source.

The present trend in the northwestern coastal belt is to establish shrimp farms in high saline areas (beyond Mundal up to Puttalam and Kalpitiya) due to unavailability of sites in low saline areas (between Chilaw and Mundal) for development. With the intensification of shrimp farming in high saline areas, the use of fresh water by installing tube wells has increased. Establishment of shrimp farms in high saline areas without assessing the adequacy and status of the water table may result in its depletion, leading to land depression. In addition, too much dependence on underground water may result in a much lower water table, which, in turn, may lead to serious competition for fresh water between aquaculture, agriculture, industry and domestic users and to drinking water becoming salty.

The establishment of many small-scale farms without proper environmental assessment led to the destruction of buffer zones between farms and between water fronts and farms and to interference with the irrigation and drainage systems. The latter causes annual flooding in the areas where shrimp farms are developed.

Recommendations

Disease Considerations

There is a need for a regular monitoring program to determine the health status and to recommend mitigatory measures to prevent possible disease outbreaks. Lack of proper monitoring makes it difficult to estimate economic losses caused by disease. A monitoring program should emphasize the following:

- Monitoring of each farm for changes in the main environmental parameters, such as dissolved oxygen deficit, accumulation of toxic nitrogen compounds, toxic algae, external pollution etc. This will enable collective measures to be taken to remedy any adverse changes before an outbreak of disease occurs. The monitoring may be done by farmers who are capable of doing so. NARA may monitor the small-scale farms and coordinate the entire program.
- Monitoring conditions that increase susceptibility to infectious diseases through stress or which cause acute effects.
- Monitoring the use and effects of chemotherapeutants. The use of antibiotics may result in some environmental changes, such as:
- qualitative and quantitative changes in the bacterial flora
- toxic effects on wild living organisms
- development of antibiotic resistance in bacteria
- transfer of antibiotic resistance to bacteria pathogenic to humans (Anon., 1991).

Due to the polluted nature of the main water sources used for shrimp farming, there is a great possibility for disease to spread throughout the area. Poor design and unplanned localization and siting of farms aggravate the risk, since infection through the water is a major source of disease transmission.

Therefore, improvements such as dredging of the Dutch Canal and opening the sea out-falls are recommended to improve the water exchange rates (Siriwardena and Dayaratne, 1990 - unpubl. doc.).

A risk analysis for health and disease management is recommended. The following areas should be examined:

- Localization of the farm, water resources and water quality
- The intake of live material at the farm
- Farming strategy, including the control of the production cycle and links with other farms
- Hygienic conditions
- Management routines
- Monitoring and control routines.

Environmental Considerations

There is a need for the Government of Sri Lanka to recognise the important role of aquaculture in its national economy, so that proper laws and regulations beneficial to its controlled expansion are established. One of the important strategies for the development of shrimp aquaculture is the adoption of transparent policies to avoid the social conflicts and adverse ecological impacts which are currently occuring.

Zoning plans, based on scientific investigations and analysis of physical, biological, social and economic information of the areas, especially in the new areas to be developed, would provide a rational approach to the allocation of land and water areas and the development of improved pond designs.

Freshwater resource use must be rationalized. Shrimp farms which are dependent upon fresh water for dilution should not be approved without knowing the availability and adequacy of the natural water table for competitive use without affecting the environment. Prior to approving further shrimp farms in the high saline areas, the following questions should be adequately addressed:

- What is the water quantity needed to dilute high saline water in the shrimp farms?
- Can the natural water table meet the requirement of fresh water for the shrimp farms?
- Has the water table been depleted or is it on the verge of depletion in the areas where shrimp farms depend on fresh water for the dilution of high saline water?
- What quantity of fresh water could be allowed for use to adjust the salinity in shrimp farms without creating user conflicts?

Besides employing the above measures to conserve water resources, construction of water recycling facilities in several demonstration farms to save water and reduce the impact on the environment should be encouraged. Placing a subscribed charge on water pumped from underground would create a water conservation consciousness among farmers. A "switching over" strategy should be adopted to change the species under culture to a relatively high salinity tolerant species. For example, relatively high salinity tolerant white prawn (*Penaeus indicus*) could be cultured in high saline areas instead of black tiger prawn, which prefers low saline conditions.

This may prevent or minimise the use of fresh water to dilute high saline water in shrimp farms. Moreover, it is generally recognised that high saline species are relatively more disease tolerant than are low saline species. When adopting the "switching over" strategy, the species should be carefully selected, taking into consideration the market trends. Has cultured shrimp production increased in relation to the increased culture area enough to justify approving a vast number of shrimp farms in the northwestern provincial coastal belt? The production figures show that it has not increased in relation to the increase in culture area. This cannot be totally attributed to the switch over from intensive to semi-intensive practice. It is partly because of the inability of hatchery production to meet the demand and partly because of the slowdown of growth rate of shrimp experienced, probably due to the polluted nature of the water source. Hence, to achieve effective culture, approval of land has to be scaled down, giving due consideration to actual hatchery production and pollution status of the water source.

To reduce the impact of shrimp farming, the "cluster farm" concept may be adopted. Almost all small shrimp farms are being operated without any mitigatory measures, such as the use of oxidation and sedimentation tanks to reduce the hazardous impact on the water source. Depending on the size of the land, the number of such small farms (not exceeding one ha in extent) in a particular area could vary. The "cluster farm" concept may be adopted in the following manner:

- Identify areas where small farms are in close proximity.
- Group small farms with a total extent not exceeding 2–3 ha to form a small unit, bringing in the concept of "cluster farms".
- Such clusters would have the following common structures, if desired, and if location allows such amalgamation:
- Common inlet
- Common outlet
- Common pumping station
- Common sedimentation and oxidation tank.

This approach would minimise the problem of locating inlets and outlets at close proximity and reduce the sediment and nutrient loading in the water source by utilising organising common sedimentation and oxidation tanks. Adoption of a system to monitor environmental licenses to prevent establishment of unauthorized farms without proper environmental assessments is important. There was no frequent and proper dialogue between the researcher, the developer and the extentionist. Without proper dialogue, it is difficult to obtain the feedback needed for formulating research to address the problems in the industry. Hence, to establish a proper dialogue between these groups via regular meetings of their representatives is recommended.

Health Management Strategy for a Rapidly Developing Shrimp Industry: An Indian Perspective

Traditional shrimp trapping systems have been in practice in the low-lying brackishwater areas of Kerala and West Bengal for several years. These systems are largely seasonal in nature. Autostocking aided by tides is practiced, and managerial inputs are minimal. Production from these traditional trapping systems is very low, averaging 100-500 kg/ha/crop.

India has vast potential for the development of commercial shrimp farming, being blessed with 1.2 million ha of coastal brackishwater land. Many coastal areas are rich in natural seed resources, and broodstock collection grounds are said to be ideal for commercial exploitation. The industry is basically export-oriented and has enormous potential to generate much needed foreign exchange. Short development periods, high return on investment and good international demand are positive features which should encourage the development of the industry. In addition, the industry can create significant employment in rural areas, including employment for women.

Realising this enormous potential, the Government has identified aquaculture as a "thrust area" under "extreme focus" for augmenting exports and earning much-needed foreign exchange. To encourage shrimp farming ventures, the Government has recently implemented a series of liberalization policies. Restrictions on the importation of aquaculture machinery, feed and aqua-chemicals have been cased, and import duties on specific items like feed have been completely slashed. Steps have been taken to encourage foreign equity participation in joint ventures and full repatriation of profits by foreign collaborators has been allowed. Five-year tax exemption has been given to companies which are 100% export oriented.

Present Status of the Industry

Realising the enormous potential for investment opportunities and taking advantage of the government liberalization policy, several national and international corporate houses, private companies and individual entrepreneurs have embarked on setting-up shrimp hatcheries and farms. Nearly 80 000 ha is under extensive or semi-intensive culture, and the quantity of farmed shrimp produced in 1994 was around 75 000 metric tonnes (mt). The area being brought under culture is increasing rapidly. At present India is ranked fourth in the world in cultured shrimp production.

In the last two to three years, growth in the shrimp hatchery sector has been phenomenal. Before 1993, seed availability was a major bottleneck. With the establishment of 100 state-of-the-art hatcheries with a total capacity of 4 billion seed per annum, this problem seems to be temporarily solved. Availability of good quality feed has not been a serious problem. Most feed is currently imported. Realising the potential demand that exists for shrimp feed, several Indian and foreign joint-venture companies have established feed mills. Aquaculture machinery and aquachemicals are being regularly imported.

Resource mobilization and financing for the development of shrimp farming have been relatively well organised, because shrimp farming is no longer seen as a traditional occupation but as a capital intensive industry. Public and private banks are extending loans to the industry on a priority basis. Several private companies have gone public and have raised sufficient funds through public issue of equity shares. Insurance companies are also involved, with risks associated with natural calamities and diseases being insured and crop losses protected.

The concept of satellite and franchise farming as promoted by big industrial houses is well established and is working reasonably well. In these systems, the industrial house acts as the nucleus for several hundred small- and medium-scale extensive farmers. Technical input in the form of consultancy and other inputs in the form of seed, feed, and disease and water quality management are provided. The end product is bought back by the company at the prevailing market price. In addition to satellite farming, many small companies have long-term buy back arrangements with big industrial houses and multinational companies.

The booming shrimp industry has encouraged the development of several small-scale aqua-related ancillary industries in rural areas and has created many employment opportunities for the rural poor.

In the last three to five years, the industry has grown by leaps and bounds, but still only a small percentage of the available 1.2 million ha of suitable coastal land has been brought under culture. In this short period, the industry has come face to face with many crippling problems. The repercussions of unregulated expansion are already being felt in many quarters. The prospect of rapid development and fast returns has led to haphazard and unorganised growth and the "fast buck syndrome" has overshadowed all relevant scientific and ecological considerations. Unplanned and unregulated mushrooming of farms around available water sources has led to negative impacts, such as serious ecological degradation, socio-economic problems, social tensions, mangrove destruction, isolation of villages and salination of agricultural land and freshwater wells.

The expansion of the industry and the desire of the farmers to make quick money have encouraged pseudoconsultants and quacks. The easy availability of a large quantity of inferior quality chemicals has led to an indiscriminate use of antibacterials and other chemicals. Bacterial drug resistance (Karunasagar *et al.*, 1994), tissue residues etc. are some of the serious problems which need to be addressed.

Aquaculture waste management has become a very serious issue. The majority of the farms use source water as a dumping ground for waste discharge. Raw effluents rich in organic matter and waste feed are released directly into water sources without any treatment or settlement. There are no community joint agreements on coordinated arrangements for water intake and effluent discharge. This has led to problems connected with disease transmission and bad water quality, and little attention is being given to pond drying, disinfection and waste removal between crops. Crop losses due to natural calamities like flooding and cyclones have become common. Absence of buffer zones and destruction of mangroves have further aggravated these problems. Since July 1994, two viral diseases (yellowhead disease (YHD), and possibly systemic ectodermal and mesodermal baculovirus (SEMBV)) have had disastrous effects on the booming shrimp industry, causing losses estimated at around Rs 600 crores (US$ 17.6 million). Since January 1995, a fallow period, or "crop holiday", has been observed in many farms along the east coast. In spite of observing "crop holiday" and adopting standard disinfection programs, the disease has reappeared in many of the farms where trial restocking is done. It appears that the industry will need some time before it can regain its original tempo and status. In view of the current industry crisis, there is an urgent need to develop and execute a rational health management strategy and to frame suitable legislation and guidelines for ensuring the development of sustainable shrimp farming. The shrimp farming industry in India needs to be developed on sustainable lines because there is still vast potential for growth, and significant potential for generating foreign exchange and rural employment.

Major Health Problems and Their Causes

In hatcheries, luminescent vibriosis (*Vibrio harveyi, V. splendidus, V. parahaemolyticus*), external fouling (protozoan ciliates like *Zoothamnium* and *Vorticella*, filamentous bacteria like *Leucothrix*) and larval mycosis (*Lagenidium*) have been consistently recognised as major health problems (Felix *et al.*, 1994). In some hatcheries monodon baculovirus (MBV) has been identified as a problem (Felix and Devaraj, 1993; Ramasamy *et al.*, 1995).

The majority of the pathogens encountered in hatcheries are normal components of the marine ecosystem. Opportunistic vibrios and external foulers become a serious problem in badly managed hatcheries. Excess stress, bad water quality and inefficient disinfection programs are some of the causes of hatchery problems. With regard to MBV, it is well known that most broodstock are natural carriers. MBV becomes a serious problem only in stressed hosts and in hatcheries lacking organised and efficient disinfection programs. Prior to April 1994, Indian shrimp farms did not experience any serious disease problems. Conditions such as external fouling, shell and appendage necrosis, black gills, luminescent vibriosis, systemic vibriosis, chronic vibriosis, muscle cramp etc. were regularly reported from farms along the east and west coast. These health problems were responsible for low levels of mortality in several farms but not for mass mortalities over large areas. Stressed shrimp tend to be sluggish and do not exhibit grooming behaviour. Colonization by external foulers is very common in weak and unhealthy shrimp, and severely fouled shrimp are susceptible to opportunistic secondary invaders like vibrios. The majority of health problems in Indian shrimp farms can be traced to excess stress associated with intensification, poor water quality, pond bottom deterioration due to waste accumulation, benthic algal growth and subsequent decomposition, lack of reservoir tank facilities for carrying out suitable prophylaxsis and bad quality source water.

The Current Crisis

Since July 1994 two diseases of suspected viral etiology (YHD and SEMBV?) have had disastrous effects on shrimp farming which has upset the industry (Shankar and Mohan, 1994; Mohan and Shankar, 1995). YHD first occurred in several farms located on either side of Kandaleru Creek, in Gudur, Andhra Pradesh, along the east coast of India. All farms in this region were receiving and discharging water into Kandaleru Creek; none had reservoirs or sedimentation tanks. The disease affected only tiger shrimp (*Penaeus monodon*) (40-70 days old) in grow-out ponds. White shrimp (*P. vannamei*) stocked separately or together with tiger shrimp were not affected. Affected shrimp did not feed and had empty guts. At first, a large number of moribund shrimp were found near the pond margin and surface. The hepatopancreas of affected shrimp was reddish yellow or pale yellow in colour and overall, the cephalothorax was yellow. Mass mortalities occurred within two to three days of the appearance of one or all of these clinical signs. To save the crop, many farmers resorted to emergency harvest.

In clinical signs, nature of mortality, species and size affected and, importantly, yellow colouration of the cephalothorax, the disease was very similar to yellowhead disease (YHD) caused by yellowhead baculovirus (YHBV). Confirmatory diagnostic reports are still not available. The second disease, now recognised as white spot disease, affected the majority of farms in Andhra Pradesh and Tamil Nadu between November 1994 and January 1995. According to latest information, it is still spreading along the east coast to many parts of Orissa and West Bengal. White spot disease affected all size groups of both tiger and white shrimp. Affected shrimp did not feed and had empty guts. Mortalities were low during the initial two to three days. A large number of circular white patches or spots were seen, first in the under surface of the cephalothorax and later in the carapace over the abdominal region. Mass moralities occurred within seven to 10 days of the first clinical signs.

In clinical signs, nature of mortality, and species and size affected, the disease is similar to IHHN caused by IHHNV. However, recently published work suggests that white spot disease is caused by systemic ectodermal and mesodermal baculovirus (SEMBV), a virus which has been responsible for the collapse of the industry in many parts of East and Southeast Asia (Nash, 1995). It is estimated that these two diseases have caused a loss of nearly Rs 600 crores (US$ 17.6 million) to the industry. The Marine Products Export Development Authority (MPEDA), concerned research organisations and experts from the Asia-Pacific Region have all recommended "crop holiday", waste removal from ponds, pond drying and disinfection programs to prevent recurrence of the disease.

Institutions and Their Activities

The shrimp farming industry has grown very fast and in a haphazard manner. This rapid growth is one of the reasons for the lack of guidelines, feasibility studies, research input etc. to the industry. Below is a brief account of the various institutions involved in the industry and their activities.

Marine Products Export Development Authority (MPEDA)

MPEDA has been instrumental in promoting shrimp culture in India, setting up model state-of-the-art hatcheries and demonstration farms in many parts of India, and has successfully demonstrated the techno-economic viability of commercial shrimp farming. In addition, MPEDA organises annual INDAQUA shows, workshops, training programs, expert consultations etc., involving farmers, industry and experts from India and abroad. As a part of the promotional program, MPEDA is also extending subsidy facilities to shrimp farming and participating in the equity of public limited companies.

MPEDA is actively associated with the ministry in farming policy guidelines and regulations relating to aquaculture. It can take the role of a nodal agency to implement policy regulations in different shrimp farming states.

Brackishwater Fish Farmer's Development Agency (BFDA)

Many of the maritime states engaged in shrimp culture have established BFDAs at the district level. Normally officers from the respective state fisheries departments are assigned to look after the activities of these agencies. These agencies are involved in identifying and earmarking government land under their jurisdiction suitable for shrimp culture and are responsible for leasing out this land to unemployed youth, people from the poor sections of society, individual entrepreneurs and companies. Organisation of a large number of promotional programs, workshops, training programs etc. is also undertaken. In many states, BFDAs have successfully established model hatcheries and farms to demonstrate their techno-economic viability. Subsidy for shrimp farming originating from the individual state governments is routed to the farmers through BFDAs.

Fisheries Colleges and Research Centres

Teaching and research institutions coming under the university of Agricultural Sciences of Karnataka, Tamil Nadu, Kerala, Andhra Pradesh, Maharashtra, Orissa and Gujarat are very closely associated with the activities of the rapidly developing shrimp industry of India. Fisheries colleges have been very successful in providing much needed technical manpower to the industry. These institutes and colleges are currently engaged in research in areas like shrimp nutrition, water quality, disease etc. Extension and training programs, and workshops are also routinely undertaken.

Central Fisheries Research Institutes

Several central research institutes *i.e.*, Central Institute of Brackishwater Aquaculture (CIBA), Central Marine Fisheries Research Institute (CMFRI) and Central Institute of Fisheries Education (CIFE), are involved in research and extension activities and are closely associated with the shrimp industry. They also conduct workshops, training programs etc. for the benefit of farmers, technical personnel and the industry.

Non-Governmental Organisations

Several NGOs have been recently established. The Aquaculture Foundation of India (AFI), Karnataka Shrimp Farmer's Forum and the Karnataka Aquaculture Society are some of the leading organisations which are actively involved in promoting interaction between farmers, evolving joint agreements on water and waste management, and conducting training programs and workshops.

Big Industrial Houses

Many of the big industrial houses have established their own research and development (R&D) facilities. Some offer diagnostic services, organise workshops and training programs, publish extension literature etc.

Banks and Financial Institutions

Banks and financial institutions are heavily involved in extending loans for setting-up hatcheries, farms, feed mills, infrastructure, R&D laboratories etc.

Insurance Sector

Several insurance companies are involved in insuring against crop losses due to natural calamities and diseases.

International Organisations

International organisations like INFOFISH and the Asian Shrimp Culture Council (ASCC) disseminate a lot of useful information on feed, water quality, disease management etc. through newsletters. In addition, these organisations also conduct training programs and workshops involving experts from India and abroad for the benefit of the industry.

Funding Support

Until recently, funding support to shrimp disease research was very limited. The recent emergence of two deadly viral diseases and the trail of destruction that they have left behind have highlighted the need for systematic and coordinated research on shrimp diseases and their management. National and international funding agencies have now recognised the need to fund research so as to develop sustainable shrimp farming in India. Several research projects are in the pipeline and many more are likely to be funded in the near future. Funding support and coordination of research effort are needed within the country.

The following are some of the national and international agencies which have funded or are likely to fund shrimp disease research in India:

- Indian Council of Agricultural Sciences (ICAR)
- Department of Biotechnology (DBT), New Delhi
- Department of Science and Technology (DST), New Delhi
- Department of Ocean Development (DOD), New Delhi
- National Bank of Agriculture and Rural Development (NABARD)
- Marine Products Export Development Agency (MPEDA)
- Various state fisheries departments
- Corporate houses
- International Foundation for Science, Sweden
- World Bank
- Australian Centre for International Agricultural Research (ACIAR).

Achievements

Indian shrimp farming is still in its infancy. The research effort that has gone into shrimp disease studies has been very limited and hence very limited progress has been made towards health management. The majority of the health management techniques and chemotherapeutic practices that are being followed are largely adopted from Southeast Asian countries.

The marketing strategy adopted by the international feed suppliers has helped in the acceptance and implementation of several health management strategies. Some of the noteworthy developments that have taken place in the last three to five years in the Indian shrimp industry are discussed below.

Recurrent problems associated with luminescent vibriosis, external foulers, larval mycosis and MBV are being effectively managed in many hatcheries. The current practice of taking good quality sea water from bore holes sunk in the sea bed near the coast has kept many problems at bay.

The use of pressurized sand filters, UV irradiation systems, better disinfection programs, and the effective use of antibacterials and other chemotherapeutants have enabled the hatcheries to function successfully in recent years. Chemotherapy for secondary and systemic vibriosis and external foulers in grow-out farms has been successfully adopted and standardized. Having realised the importance of health management for successful aquafarming, the Government of Karnataka has established a Disease Diagnosis Centre at the College of Fisheries, Mangalore. This centre has a mandate to offer disease diagnostic and health management services to the farmers of the state. Several maritime states are considering establishing such centres.

Several research projects on shrimp diseases have been initiated. Some of the priority areas that have been identified are development of immunodiagnostics for screening and rapid diagnosis, development of vaccines and immunopotentiators, shrimp virology and evaluation of routine health monitoring programs. Over the last two years, a large number of training programs and workshops involving experts from India and abroad have been organised. These training programs and the current crisis the industry is facing has certainly impressed upon farmers the need for effective health management strategies and proper regulations for sustainable growth of the industry.

Many public limited companies have successfully established R&D centres. These centres will play a vital role in implementing rational health management practices over large areas through satellite and franchise farming.

In view of the current crisis, the Government of India has decided to issue aquaculture guidelines to all the concerned states to ensure the safety and sustainability of all aquafarming operations. Earmarking specific areas, statutory "crop holiday" or crop rotation, restricting maximum stocking density to < 30 m^2 and complete drying of the pond bottom for two to three months are some of the areas in which the Government has decided to issue guidelines.

Constraints in Health Management

Management Constraints

Short gestation periods and high returns on investment have overshadowed many scientific and ecological considerations during the development of the shrimp industry. Adopting effective health management has therefore been very difficult.

- Dependence on a common water source is the root cause of many problems Most farms do not have reservoir and sedimentation tank facilities. Lack of reservoir tanks makes it impossible to undertake prophylactic measures, while lack of settlement tanks for sedimentation and disinfection of waste water before discharge adds to the problem of health management. The majority of the farms are owned by small farmers (< 5–10 ha) and it will be difficult or them to make provisions for reservoir and sedimentation tanks.
- Lack of a sufficient number of disease diagnostic centres.
- Lack of a sufficient number of trained personnel for disease diagnosis.
- Failure to implement routine health monitoring programs. From the point of view of chemotherapy, the normal practice of diagnosing the cause of a disease during a mortality serves little purpose.
- The nature of the farms and the dependence on common water source make it very difficult to follow health management strategies at the individual farm level.
- Lack of coordinated joint agreements between farms on waste disposal, effluent discharge and water intake.
- Lack of awareness among farmers on the need for proper disposal of dead shrimp and waste water during and after a disease outbreak.

Constraints in Regulation

The nature and extent of the shrimp farming industry in India make it very difficult to follow standard and scientific health management strategies. Lack of adequate regulations and absence of nodal agencies to implement such regulations are serious constraints to enforcing strict health management over a large area. At the national level, there is no comprehensive aquaculture policy and there is no nodal agency to enforce regulations. Some of the important issues which need immediate attention are:

- The process of disease notification is absent in the country and hence it is difficult to restrict in-country movement of infected seed and broodstock.
- Quarantine programs for new introductions and imports are not effectively implemented.
- Disease certification programs are not practiced.
- Rapid disease screening systems are not available.
- Enforcing 100% "crop holiday", if and when required, is very difficult.
- No effective regulations exist to check the misuse of antibacterials and other therapeutic chemicals.
- Entry of pseudoconsultants and quacks has led to large-scale availability of bad quality chemicals and their misuse.
- Lack of effective regulation and its implementation to halt the ecological degradation associated with unplanned mushrooming of farms around available water sources makes the problem of health management worse.

Recommendations

The Indian shrimp industry is still at a very early stage of development. The industry has enormous potential to generate much needed foreign exchange and significant rural employment. In the interest of the nation, it is essential that it develop in a planned and regulated manner on a sound ecological and scientific basis. This can be ensured only if proper future directives are developed and effectively implemented. The future directives should be in the areas of policy and regulations, management, and research and development.

Regulations

Practical and rational regulations should be developed and implemented to guide the development of the industry. Some important areas which need to be considered are:

- Development of a comprehensive national policy on aquaculture.
- Restriction of uncontrolled shipment and importation of broodstock and larvae.
- Implementation of organised quarantine, eradication and health certification procedures.

- Regulation of man-made canals and creeks bringing brackishwater deep into agricultural land.
- Conversion of agricultural land to shrimp ponds.
- Regulations to check the misuse of chemicals
- Mandatory waste removal, drying and disinfection programs between crops at farms.
- Enforcement of complete "crop holiday", when required
- Restriction of the maximum stocking density to < 30 m^2

Management

Adopting some of the following management measures would go a long way in aiding the development of sustainable shrimp farming in India:

- Routine health monitoring should be mandatory for all farms. Early detection of disease would permit rapid treatment, adjustment of management or emergency harvest.
- Disease notification should be mandatory, so that restrictions can be enforced on movement of live seed and broodstock within the country.
- Routine prophylactic husbandry practices should be made compulsory to obtain licenses, bank financing, insurance coverage, subsidy etc.
- Each farm should allocate areas for reservoir and settlement tanks.
- Waste management should be strictly enforced (sedimentation tank).
- Organised disinfection programs should be implemented.
- Community agreements in water management (water intake and waste water discharge) should be developed.
- Closed/semiclosed systems of culture should be considered.
- Mangrove destruction should be prevented.

Research and Development

Research and developmental work in the following areas would help evolve standard scientific health management tools:

- Exploring the use of specific pathogen-free broodstock and healthy post-larvae.

- Exploring the development and use of virus-resistant species.
- Developing immunodiagnostic kits for rapid disease diagnosis, and screening of broodstock and post-larvae.
- Determining the carrying capacity of brackishwater sources before developing farms and earmarking areas for shrimp farming.
- Establishing disease diagnostic centres in shrimp farming belts.
- Creating a national centre of excellence in shrimp disease research.
- Developing immunoprophylactics, including vaccines, anti-stress factors, immunostimulators etc.
- Initiating work on kinetics of antibacterials in shrimp tissue, tissue residues, bacterial drug resistance etc.
- Developing bioremediation technology for shrimp farming.

6

Expression Pattern of Genes of Hormones that Control Growth in Fish

In aquaculture or the farming of fish in a controlled environment like in fishponds, in land-based tanks, and in net cages in the sea, growth is highly dependent on food availability. Whether the culture system used is extensive, where the fish stocks rely mainly on the presence of natural food like algae and other aquatic organisms in the culture environment such as in fishponds, or intensive, where the stocks are largely, if not totally dependent on artificial feeds like in high density culture of milkfish in open sea cages, the food factor is a major determinant of success. Because giving excess feeds is mistakenly equated with faster growth, fish stocks tend to be overfed, as usually is the case in intensive culture conditions. Since feed cost is a major cost item in intensive aquaculture operations constituting roughly 60% of the production cost, any means of reducing this without affecting the growth of the fish will certainly help the fish farmers.

Growth in fish, as in other mammals, is a complex process. Genetics, nutrition, environment, and many other physiological factors are involved. Several hormones in the "brain-pituitary-liver axis" or the "growth axis" play a key role in this process. These hormones include the growth hormone releasing hormone (GHRH) and the growth hormone release inhibiting factor (GHRIF) in the brain, growth hormone (GH) in the pituitary gland, and insulin-like growth factor I and II (IGF-I and IGF-II) in the liver. Upon stimulation by GHRH from the brain, the pituitary gland secretes GH into the bloodstream, and GH in turn stimulates the liver to secrete IGFs, mainly IGF-I. IGF-I then binds to its receptors in many organs in the body and this binding triggers a series of cellular events that result in processes like cell proliferation and differentiation, and ultimately results in growth.

All these hormones act in concert to effect growth and their secretion patterns follow a rhythm according to a given set of environmental and physiological factors. One of these factors is nutrition. This most recent study of ours tries to look at whether the gene expression patterns of these hormones (GH, prolactin (PRL), somatolactin (SL), IGF-I and IGF-II) change with a shift in the feeding schedule as well as during times of food restriction or starvation.

This is as a prelude to future work of developing a feeding strategy that is applicable in aquaculture systems that would reduce the amount of feeds given, thereby reducing the feed cost, without affecting the fish' growth. PRL and SL were included in the investigation since these two hormones are members of the growth hormone family of proteins. PRL and SL share structural similarities with GH and as such they also share some biological properties and have overlapping functions.

Using rabbitfish as the experimental fish, a series of experiments were done to address 2 basic questions. The first was whether the expression pattern of these important hormones changes with a shift in the feeding time. The hypothesis was, if the gene expression follows a rhythm in relation to the feeding time then an expression pattern may be observed in fed fish with feeding time as reference, and that this rhythm may be non-existent in starved fish. Three series of experiments were done to address this question with feeding time as the variable. In all 3 experiments, fish were first acclimatized to the experimental tank conditions for 3 days during which food was given in the morning at about 10:00 AM.

After the acclimatization period, the fish were reared for another 7 days during which the feeding time was set at 10:00 AM (Experiment 1), 3:00 PM (Experiment 2), and no feeding (Experiment 3). After 7 days of feeding at the prescribed feeding time (for experiments 1 and 2) and 3 days of no feeding (for experiment 3), representative samples of fish were sacrificed every 3 hours for 24 hours and the pituitary glands and the liver were removed for measurement of the gene expression levels of GH, PRL and SL, and IGF-I and II, respectively.

No pattern was observed in GH gene expression in relation to the feeding time. However, a diurnal pattern with high GH levels during night time and low levels during day time is observed which is similar in other mammals like in humans. Gene expressions for PRL and SL were also irregular and did not follow a pattern with feeding time. SL gene expression, were low during day time especially during mid day which is consistent with the established inverse relationship between SL's physiological activity and illumination levels in fish.

While the expression of the three pituitary hormones do not seem to follow a pattern with feeding time, IGF-I gene expression in the liver showed a trend. IGF-I gene expression pattern was influenced by feeding time where the highest levels were recorded consistently 5-6 hours after feeding.

This pattern in IGF-I gene expression was not observed in the group of fish that were not fed. This observation suggests a particular need for IGF-I some time after feeding probably to aid or hasten the chemical breakdown of food by inducing the secretion of key digestive enzymes and facilitate the absorption of nutrients. Indeed, there is evidence that IGF-I can induce the activity of lipase, an enzyme that catalyses the breakdown of lipids (Tremblay et al., 2001).

While IGF-I gene expression consistently showed a rhythm with feeding, IGF-II gene expression was irregular and did not follow a pattern with feeding time. A common observation in all 3 experiments, however, is the consistently high level of IGF-II during early morning (06:00 AM). The physiological significance of this high IGF-II level early in the morning is presently unknown, although it may be related to the onset of light as the photic cue that synchronizes the daily secretion patterns of some hormones as observed in other fish species like rainbow trout (Boujard and Leatherland, 1992).

The second question was whether the hormone expression pattern changes when food is limiting or during periods of short-term starvation. A 15-day starvation experiment was conducted. Two groups of fish were acclimatized for 7 days in experimental tank conditions and feed daily at 10:00 AM. On the 8th day, food was withheld in one group and this continued for another 15 days (starved group). After 15 days of no feeding, the starved group was re-fed for 6 days.

The control group was given food for the entire duration of the experiment. In both groups, representative samples of fish were sacrificed at the start of the experiment (for initial samples), at days 3, 6, 9, 12, and 15 of starvation, and at days 3 and 6 of re-feeding. During every sampling, the pituitary glands and livers were removed for the determination of gene expression levels of GH, PRL and SL in the pituitary gland, and IGF-I and II in the liver.

GH gene expression was significantly elevated starting on the 9th day of starvation and remained high until the last day of starvation. The levels returned back to normal when the starved fish were re-fed, a clear indication of GH response during times when food is limiting. Although metabolism is one of the many functions of PRL, very little is known about its involvement during fasting or starvation in fish. In our study, a very clear reduction in PRL gene expression was observed with starvation.

The level started to decrease on the 6th day of starvation and it became more pronounced as starvation progressed. Interestingly, the level returned back to normal after re-feeding. Our results are the first report on a very clear reduction of PRL gene expression during starvation in fish.

This finding about PRL is not totally surprising since in mammals PRL is known to affect metabolic homeostasis by regulating key enzymes and transporters that are associated with glucose and lipid metabolism (Ben-Jonathan et al., 2006). Since there is a need to mobilize the fat reserves to supply the energy requirement during period of starvation, the enzymes involved in fatty acid mobilization and synthesis need to be activated.

Because of the inhibitory role of PRL on these enzymes, it makes sense that during period of energy deficit such as during starvation, PRL level will be reduced. In the case of SL, there is some evidence for its involvement in energy homeostasis in sea bream. Like in other fish species studied to date, however, we did not observe significant changes in the expression of SL genes in fish that were fed and starved. This may indicate that unlike GH and PRL, SL may not be critically involved or play an active metabolic role during starvation in fish.

IGF-I, like GH, have metabolic functions so that it can only be expected that nutritional status is one of the factors that regulate GH-IGF-I axis in fish. The typical response of fish to fasting or starvation is an increase in both GH gene expression and GH protein level in the blood, and a decrease in liver IGF-I gene expression and IGF-I protein level in the blood.

This opposing picture of GH and IGF-I activity during fasting is explained by the development of GH resistance in the liver which results in the decrease in the number of GH receptors that consequently leads to reduced GH binding and impairment of the GH signalling pathway. The results of the present study is consistent with what is observed in many fish species, except that a significant increase in IGF-I gene expression was observed early in the starvation period, during days 3 and 6 after starvation, which was not observed in many previous studies.

Like IGF-I, IGF-II is also a potent growth factor especially during early development. It stimulates cell proliferation and DNA synthesis in zebrafish embryonic cells (Pozios et al., 2001). Whether it also functions as a metabolic hormone is not yet known although there is preliminary evidence that IGF-II regulates metabolism in trout muscle cells (Codina et al., 2004). Unlike IGF-I, however, where a very clear relationship with GH exists during starvation, nothing is known about the behaviour of IGF-II during altered nutritional status in fish.

In higher vertebrates, though, there is evidence that IGF-II is regulated by nutritional status and is age dependent as in sheep where it is more sensitive to nutritional status in older but not in younger sheep (Hua et al., 1995). In the present study, the initial response of IGF-II to starvation is an increase in its gene expression which occurred after 3 and 6 days of starvation after which the levels returned back to normal and were no longer different from the fed group.

As expected, the body weight of the fish in the starved group decreased during the 15 day starvation period but the weight reduction was not significant - a mean weight loss of 3.8 g during 15 days. The fed group continued to grow during 15 days experiment gaining a mean body weight gain of 3.7 g. What is interesting is that the starved group showed tendency to catch up growth after re-feeding which in our experiment was only for 6 days.

There is some evidence of catch up growth in fish after re-feeding following some period of food restriction (Xie et al., 2001). As shown also in the results, the expression of GH and IGF-I, the major hormones that control growth, were also normalized soon after re-feeding. Considering that the fish will exhibit compensatory growth when re-fed after some period of food restriction, there exists the potential of subjecting the fish to some degree of feed restriction during the culture period.

A reduction in the amount of food will be realised during the feed restriction period and yet not affecting the fish' overall growth. The appropriate duration of feed restriction is critical in this regard. Once established, this promises to be a good approach to reduce feed cost without affecting total production.

Prevention and Control of Fish Diseases

Role of Stress in Fish Diseases

Stress is debilitating to fishes and greatly increases their susceptibility to various diseases, as shown below according to Wedemeyer and McLeay (1981):

Disease	*Environmental Stress Factors Predisposing to Disease*
Furunculosis (Aeromonas	Low oxygen (4 mg 1^{-1}); crowding; handling in the presence
salmonicida)	of A. salmonicida; handling for up to a month prior to an expected epizootic
Bacterial gill disease	Crowding; unfavourable environmental conditions such
(Myxobacteria spp.)	as chronic low oxygen (4 mg 1^{-1}); elevated ammonia (0.02 mg 1^{-1} unionized); particulate matter in water
Columnaris (Flexibacter	Crowding or handling during warm (15°C) water periods
columnaris	if carrier fish are present in the water supply; temperature increase to about 30°C, if the pathogen is present, even if not crowded or handled
Kidney disease	Water hardness less than about 100 mg 1^{-1} (as $CaCO_3$);
(Renibacterium salmoninarum)	diets containing corn gluten or of less than about 30% moisture
Haemorrhagic septicaemia	Pre-existing protozoan infestations such as Costia,
(Aeromonas and	Trichodina; inadequate cleaning leading to increased
Pseudomonas spp.)	bacterial load in water; particulate matter in water; handling; crowding; low oxygen; chronic sublethal exposure to heavy metals, pesticides or poly-chlorinated biphenyls (PCBs); for carp, handling after overwintering at low temperatures
Vibriosis (Vibrio anguillarum)	Handling; dissolved oxygen lower than about 6 mg 1^{-1}, especially at water temperatures of 10–15°C; brackishwater, of 10–15‰ salinity

Contd...

Parasite infestations (Costia,	Overcrowding of fry and fingerlings; low oxygen; excessive
Trichodina, Hexamita)	size variation among fish in ponds
Spring viremia of carp	Handling after overwintering at low temperatures
Fin and tail rot	Crowding; improper temperatures; nutritional imbalances; chronic sublethal exposure to PCBs; or to suspended solids at 200–300 mg l^{-1}
Coagulated yolk of eggs	Rough handling; malachite green containing more than
and fry	0.08% zinc, gas supersaturation of 103% or more; mineral deficiency in incubation water
"Hauling loss" (delayed	Hauling, stocking, handling in soft water (less than 100
mortality)	mg l^{-1} total hardness); mineral additions not used; CO_2 above 20 mg l^{-1}
Blue sac disease of eggs	Crowding; accumulation of nitrogenous metabolic wastes due to inadequate flow patterns

Prevention of Diseases

Prevention, rather than treatment, should be the aim of the fish culturist. Good management of fish farms is of primary importance in avoiding disease and parasite problems. The following general principles should be followed.

Provision of Pathogen-Free Water

When pathogen and germ-free water from springs or wells is used to supply fish farms, fish can be reared under parasite-free conditions for a relatively long time. Such water is essential for hatcheries, and it is also necessary that fry rearing ponds obtain water of the best possible quality. Ideally, water should not flow from one pond to another, though due to water shortage this rule cannot often be fully observed. As second-best, water flow should be from ponds containing young, less-infected fry towards those with older, potentially more infected fish.

Control of Wild Fish

Wild fish living in the canals and ponds of fish farms are hosts and vectors for pathogens. Their access can be hindered by placing wire-mesh screens over the water inflows to ponds. If necessary, chemicals can be applied to eradicate them. Deliberate introduction of fish from other hatcheries should preferably be avoided. When this is not possible, the fish should be obtained from a hatchery which has no history of serious disease problems; and also, be treated for ectoparasites prior to stocking.

Stocking Density

Avoid overcrowding fish at any time, and particularly during hot weather.

Water Quality

Fish pond and hatchery water must be provided in sufficient quantity, at a suitable temperature, free of pollutants, and rich in oxygen.

Age Segregation

Fish of different age groups must never be stocked together in the same pond. Older fish are generally infested by parasites, which they can easily transmit to young individuals. Similarly, so-called "after stocking" is bad practice, since fingerlings stocked 2–3 weeks after the initial stocking of the pond can become heavily infested by the older fish, sometimes resulting in massive losses. Where artificial hatching is practised, fry can develop under parasite-free conditions for longer. If natural spawning and hatching is inevitable, preventive treatment of breeders against parasites is of paramount importance, as is their removal from the pond as soon as possible after spawning.

Drying, Freezing and Disinfection of Pond Bottoms

Most parasites develop in annual cycles. Their eggs or spores survive in the soil of pond bottoms. Periodic drying of a pond kills these eggs and spores, as well as their intermediate hosts (e.g., snails), thus interrupting the parasite's life cycle. In intensively managed fish farms, occasional drying of ponds is essential. In ponds dried and subjected to freezing, even the most resistant parasites are killed, and the effect of long-lasting drying of ponds under tropical conditions is almost the same. In each case the disinfection of any remaining pools of water with lime is necessary. If there is no possibility for drying, the whole pond bottom should be disinfected with 2.5 t/ha of lime.

Preventive Elimination of Parasites from Fishes

Transferring fish into a bigger pond itself decreases the chance for parasites to reproduce. Bath treatment applied to fish before stocking into a new pond improves the situation further.

Control and Therapy of Fish Diseases

Even on fish farms where management practices and feeding standards are good, disease and parasite problems can occur from time to time, and control measures become necessary. According to the type of pathogen, culture system used, degree of incidence and intensity of infection, and other circumstances, one or more of the following control measures may be appropriate.

Test and Slaughter

Where tests indicate the presence in the fish of an incurable, virulent infectious agent, slaughter of the entire stock is sometimes the only way of eliminating the pathogen from the farm and ensure it does not spread to other units. This extreme remedy is most often necessary in cases of viral infection. Some countries have laws requiring compulsory slaughter of stock when certain diseases are diagnosed, though most of them do not provide the financial compensation which would be payable to farmers of terrestrial livestock in the same circumstances.

During slaughter, all ponds, tanks, etc., must be thoroughly sterilized before re-stocking with new fish. Often fish slaughtered because of disease problems must be disposed of in lime pits or otherwise hygienically destroyed. However, for certain diseases where spread from dead carcasses is extremely unlikely, and when no public health risks are involved, slaughtered fish can be sold on the market in the normal way.

Quarantine and Restriction of Movement

Quarantine is when fish which are to be moved from a suspected or infected geographical area to a non-infected geographical area must be held in detention for some time, at least as long as the incubation period of the suspected disease. They can then be moved into the new geographical area only if the suspected disease does not develop. Restriction of movement is when all transfers of fish from infected to non-infected areas are prohibited. Frequently, sales of fish to the market are also forbidden. Quarantine and restriction of movements can be used successfully only through a good cooperation between fish culturists, fish health specialists, state agencies and everyone interested in controlling fish diseases.

The most effective use of quarantine and movement restriction for disease control has been through legislative regulation of the intercontinental, interstate or interprovincial movement of fish.

Quarantine and restriction of movement have been found to be especially effective when combined with test and slaughter or sanitation and disinfection.

Immunization and Disease Resistance

Immunization and disease resistance have been of limited use for control of infectious diseases of fish. The reasons why fish immunization has been of limited success are:

a. Fish are not as immunologically competent as higher animals, especially at lower temperatures; and
b. The technology for mass immunization of cultured fish is limited. Injection of all fish in a fish culture facility is usually impossible.

The oral method of immunizing large numbers of fish against bacterial pathogens has been of limited use. The procedure used for oral immunization has been to add the immunizing agent to fish food and to feed at prescribed time intervals. Mass immunizing techniques in which fish are first submersed in a hyperosmotic solution followed by immersion in a hypoosmotic bacterin are more adaptable to the requirements of the large numbers of fish in most culture facilities. Fish are placed in a bacterin which is contained in a pressure vessel. A partial vacuum is established inside the vessel and then released. The rapid change in pressure causes some of the bacterin to enter tissues of the fish. This method has been used experimentally but holds little promise for mass immunization of fish. Immunization of large numbers of fish has been accomplished by placing them into water suspensions of immunizing agents and allowing direct absorption.

Spray methods may also be used by spraying the immunizing agents against the fish. These two methods of mass immunizing of fish are being developed rapidly because both are less traumatic to the animals. Resistant strains of fish species have been used to control disease. Attempts have been made to develop strains of fish which are resistant or immune to certain pathogens. This method of disease control holds promise but requires much more research to be effective.

Drug Therapy and Disinfection

Drug therapy and sanitation is the method of disease control which normally comes to mind whenever treatment of a disease is mentioned. Therapeutic drugs have been found to be effective in the control of certain diseases of humans and domestic animals. However, drugs used for systematic therapy, disease prophylaxis and disinfection procedures have been of limited use for diseases of fish. There are an extremely large number of known therapeutic compounds available from pharmaceutical manufacturers which have not been applied to the control of fish diseases. There are a great number of synthetic compounds prepared by chemists and pharmacologists each year which may be potentially effective in the control of pathogenic organisms. Only a very limited number of all of these compounds have been examined for possible usefulness as therapeutic compounds for fish disease control. Those compounds which have been tested and licensed for use with fish can conveniently be considered according to their method of application, as follows: Individual treatments (e.g., injections). Except for very valuable breeders, treatment of individual fish is not normally feasible or economic. Normally mass, or stock, treatment is applied in fish farms both for preventive and therapeutic purposes. Medicaments against parasites can be applied in the form of baths or in medicated feeds. However, intraperitoneal injection can be very effective against threadworms. Bath treatments. External parasites can best be eliminated by treatment with antiparasitic medicines dissolved in water. Three types of bath treatments can be given.

Short bath: Infested fish are placed into the bathing solution for between one minute and one hour. For this purpose, tanks made of wood or synthetic material are suitable. Metal tanks are less satisfactory, since poisonous components can dissolve from their surfaces during the treatment. For treatment in ponds, different dilutions of common salt and formalin are used.

A 15-minute bath in a 2.5% salt solution can be used against various unicellular parasites, fish lice and leeches. This cheap procedure, based on the effects of osmotic differences, is easy to apply and, though it does not give perfect results, it meets the requirements of practical purposes. Bathing in 1:5 000 formalin for 45 minutes is effective against unicellular parasites, e.g., Chilodonella and Costia. Long bath: Very dilute solutions of malachite green or organic phosphate esters are used for treatment in ponds. Since water exchange is slow, contact time will be long.

Malachite green (zinc-free), which was originally used as a staining material, is now widely employed against unicellular parasites. It is the only efficient chemical against "ich" disease. Malachite green can be used at a concentration of 0.1–0.2 mg/l for 24–48 hours in small ponds. The required amount of chemical can be dissolved in lukewarm water and poured in at the pond inflow. Inflow water subsequently carries the chemicals to every part of the pond. The outlet sluice must be closed when the green colour can be seen there. For common carp, concentrations as high as 0.4–0.8 mg/l can be used. However, herbivores, catfish and eels, are more sensitive to this chemical, and concentrations of only 0.1–0.2 mg/l are recommended for them. Since the toxicity of malachite green varies between batches, it should be tested on a few fish before widespread use.

A 24-hour bath will free fish of Chilodonella and Trichodina, while a 48-hour treatment also kills all the developmental stages of Ichthyophthirius multifiliis. Malachite green is of no use against metazoans and spore-bearing protozoans. Various copper compounds are used in cases of fungal infestation. For the disinfection of water, copper sulphate is applied at the rate of 1–2 kg/ha (where the water column is 1–2 m); malachite green can be used to eliminate fungi from eggs and fish. Because the zinc content of malachite green can be toxic, test bathing should be done before the full treatment.

During decomposition, dilute malachite green solution becomes toxic. Bathing solutions should therefore always be freshly prepared. This also holds true for most of the other compounds used for bath treatments. Due to its intensive and lasting staining effect, malachite green must be applied with much care. Fish bathed in this chemical will retain the colour for 6–8 days, and should not be consumed until after that period. Organic phosphate esters: Low concentrations of various derivatives of these chemicals (Ditrifon, Flibol, Neguvon, Dipterex, Masoten) are extremely effective against Dactylogyrus, leeches, fish lice and Lernaea when applied as a bath for 24–48 hours. Most of the parasites die within 24 hours. One advantage of these chemicals is that, at the low concentrations used, they decompose very quickly in the pond. It is therefore not necessary to change the water after treatment. Their disadvantage, however, is that they also damage food organisms. Consequently, increased attention must subsequently be paid to proper feeding.

In-transit bath: This is a very practical way to eliminate parasites while fish are being transported from one place to another. Organic phosphate esters are especially suitable for this; during a 0.5–2 hour-long transport, a 0.1 g/1 Ditrifon solution will free fish of monogeneans and parasitic copepods. Protozoans can also be eliminated in a similar way with appropriate concentrations of formalin or malachite green. Medicated feeds. Parasitic diseases can be cured with medicines mixed into feed or with premixes. This type of therapy is especially effective against tapeworms, but is also suitable for the elimination of nematodes and coccidia. For the treatment of Bothriocephalosis in carp farms, feed containing 0.1–0.2% Devermin can be fed for 1–3 days.

In bacterial diseases, Furane derivatives (furazolidone, nitrofurazone, sulphonamides), sometimes in combination with trimethoprin, or antibiotics (oxytetracycline, chloramphenicol, neomycin), are applied with feed. Before treatment begins, tests for drug resistance should be done. In applying these drugs, the recommended dosage and period of treatment should be respected, so as to prevent the development of bacterial resistance. For myxobacterial infections, the drugs are routinely administered mixed with feed, only rarely dissolved in water.

Disruption of Parasite Life Cycles

Destruction or reduction of a link in the transmission cycle has been used to control infectious diseases of fish involving animal parasites. Metazoan parasites of fish have a definite transmission cycle involving the fish at some stage of development. Many of these parasites require one or more other animal host species to complete the life cycle. Each stage of development in each host offers a possible means of disrupting the transmission of the parasite. However, in practice eliminating a link in the transmission cycle may not be feasible, because it may mean the elimination of a protected mammal or bird, or of a crustacean or mollusc which can be difficult to eliminate.

Importance of Fish Food Quality

Food has a decisive role in fish farming. On surveying the different fish farming operations, from the extensive ways up to the intensive systems, feed will show an increasing importance. Consequently the risk of diseases caused by food quality problems or by pollution of water also increases.

Impact of Food Quality on the Aquatic Environment

This problem is of most significant importance in intensive fish rearing systems where biologically complete feeds are used. Certain compounds - e.g., nitrogen, phosphorus, protein - in these feeds will dissolve in water if not properly stabilized, thereby increasing its oxygen uptake and ammonia level. This also favours the propagation of several harmful bacterial and algal species in fish rearing systems. It is thus important to reduce the disintegration of these mixed feeds to a minimum. Fats and carbohydrates in the diet not only provide concentrated metabolizable energy, but stop the compounds dissolving and reduce phosphorus content to less than 7 g/kg.

Relationship between the Quality of Food and the General Resistance of Fish to Diseases

Compounds of the food absorbed from the intestine pass through the liver. There they become utilisable for the organism, are detoxified, and sometimes stored. During this process serum proteins important for the organism, and other biologically active compounds ensuring a general resistance to diseases, will also develop here. Therefore any damage to the liver, including those caused by the food, lead to a decrease in quantity of these compounds and to impairment of the general resistance of the organism. This favours the rapid growth of saprophytes and facultative pathogens, and thus to development of disease.

Deficiency Diseases

Deficiency diseases are encountered predominantly in intensive rearing systems where natural food is not available and the fish are reared entirely on pelleted feed. Deficiency diseases caused by lack of inorganic micro- and macro-elements are rarely found even under intensive conditions, because fish can take them from the water or the feed. However, iodine deficiency has occasionally been observed in trout-fry rearing. Genuine deficiency diseases are most often caused by lack of vitamins. Most experience has been gained in trout farming.

Diseases Caused by Toxic Compounds in the Feed

Unfortunately it is not uncommon in many parts of the world to find feeds which have been spoiled by microbial action or polluted with chemicals or pesticides being fed to fish, in the mistaken belief that they can be fed without any risk under semi-intensive conditions. On the contrary, however, these polluted feeds cause serious inflammation of intestines, liver damage and sometimes fish kills. In addition, they cause poor appetite, impaired growth, and retard weight gain of fish for 2–3 weeks. Thus a few such episodes during a growing season can cause irreparable damage to productivity.

Fats become rancid through the oxidation of unsaturated fatty acids. In the process Vitamins A and E are decomposed and damage to the liver and genital tissues occurs. Cereal grains treated with caustics (e.g., mercury TMTD) are dangerous since the toxic agents accumulate in the fish and might be harmful to humans. Fungi such as Aspergilus and Fusarium produce toxins such as aflatoxin, T_2 toxins and ochratoxin. These toxins attack various organs of the fish when they enter it with polluted feed. Aflatoxin causes liver cancer in trout, F_2 toxin impairs the reproductive capacity of carp, inhibiting the genital production. T_2 toxin causes poor appetite, impaired feed conversion ratio, and prolonged exposure injures the immunity system. The above problems can be avoided by using only feeds of the best possible quality.

Fatty liver in trout can be the result of feeding diets of high carbohydrate content. If protein content of feed is lower than 25%, the development of the disease is very likely, but is also made worse by Vitamin E deficiency. The disease manifests itself in the accumulation of fat in the liver and at the same time in a dramatic decrease of glycogen. Accumulation of fat is followed by lesions in liver cells. Lesions become infiltrated by lymphocytes. The colour of the liver becomes light, resembling that of a fat goose. Owing to the damage to the liver, the general resistance of fish decreases and they can sometimes die of quite simple causes. The disease cannot be cured, but by feeding diets of 40–50% protein content it can be prevented.

Diagnosis of Diseases Caused by Feed

Diagnosis of these fish diseases is a very difficult task since such damage very rarely manifests itself in a clear form. Since the development of the disease is accompanied by a decrease in general resistance, common/routine diseases can appear (external parasites, facultative pathogenic bacteria, etc.). Diagnosis of these diseases needs wide-ranging testing. First, environmental effects, primary infections and parasitic infestations, have to be ruled out. If this has been reliably done, feed-caused disease can be suspected. The presumptions concerning the cause of the disease can be confirmed by changing feed (after feeding with good quality and biologically complete feed, the fish should recover). Generally in such cases the disease can also be experimentally induced.

Environmental Toxicosis

Fish are the best indicators of environmental pollution. All substances which the fish encounters during its lifetime can be detected in its tissues. Consequently, the examination of fish body composition will give a reliable picture of the level of pollution in the fish pond. Pollution in fish ponds is a result of agricultural and industrial activities. In agriculture, inorganic fertilizers (those with high ammonia content are the most dangerous) and pesticides (chlorinated hydrocarbons, organic phosphate esters, carbonates) are responsible for pollution. The most dangerous pollutants of industrial origin are heavy metals, raw oil derivatives, and chlorine.

Heavy pollution of the environment can cause toxicosis and fish kill. If a high number of fish of mixed ages and species die simultaneously, toxicosis can be suspected as the probable cause. However, before starting any investigation, the possibility of oxygen deficiency should first be ruled out. This occurs either when fish cannot take up sufficient oxygen due to a gill problem, or because the oxygen level in the water is very low. Oxygen deficiency may occur if a high amount of organic material is dumped into the water (generally from pig farms, slaughterhouses, sugar processing or distillery plants). The decomposition process of this organic material uses up oxygen from the water.

Some of the above-mentioned toxic agents (ammonia, hydrogen sulphide) may be released during the decomposition of organic material, while others are carried in with the inflow water. Lime and inorganic fertilizers are washed into the water by rain and snow. Active substances in insecticides (organic phosphate esters, chlorinated hydrocarbons) enter the water either from the fields or as a result of negligence during spraying.

These substances are neurotoxins, and the symptoms of this type of toxicosis in fish are writhing and convulsive swimming. Generally, the cause of death is respiratory paralysis. No characteristic symptoms are produced by industrial pollutants (detergents, heavy metals, etc.). No firm diagnosis can be made from the symptoms alone, and even histological examination cannot give conclusive information. Extensive investigations covering every possible parameter are necessary to establish the cause of toxicosis. During a mass fish kill bacterial diseases, parasitic infestations and oxygen deficiency can be reliably ruled out. However, toxicosis can be the cause even in cases where negative toxicological results are obtained. Examination of toxicosis needs great care and circumspection, since heavy fish losses are often followed by legal action to establish liability.

Natural and Artificial Gynogenesis of Fish

Gynogenesis is a special form of sexual reproduction in which insemination is necessary but the head of the sperm penetrating into the ovum does not transform into male pronucleus; and the gynogenetic embryo develops at the expense of the ovum nucleus only. Consequently the gynogenetic offspring are all females identical to the mother. Reproduction of gynogenetic forms takes place when gynogenetic females mate with males of the bisexual form of the same and related species.

Gynogenesis cannot be regarded as usual fecundation which is characterised by amphimixis. It is not identical with parthenogenesis either, though very close to it. Brachet (1917) wrote that gynogenesis was a bridge built by nature connecting fecundation with natural parthenogenesis. Gynogenesis was first found in some species of free-living nematodes. It is now known to exist also in some other worms, insects, fish and amphibians. Most cases of natural gynogenesis were discovered in the past ten years. Natural gynogenesis occurs more often than believed, though compared to parthenogenesis it is extremely rare. Two cases of natural gynogenesis are known in fish, namely in Carassius auratus (Cyprinidae) and in live-bearing Mollienesia formosa (Cyprinodontidae); of which the gynogenesis of C. auratus has been studied more thoroughly because of its importance in the artificial rearing of this species.

Natural Gynogenesis of Carassius Auratus Gibelio

The geographical distribution of C. auratus extends from Japan and China in the East to the countries of West Europe (Poland, Romania, Germany). The waters in Japan, China, Korea and the adjacent countries are populated by the typical form of this species, C. auratus L., and in areas farther west by its subspecies, C. auratus gibelio Bloch. Detailed information on the systematic status and biology of C. auratus can be found in the works of Berg, (1949) and Nikolsky (1954). In nature C. auratus gibelio (and evidently C. auratus L.) occurs in two forms, the usual bisexual form having females and males and the unisexual gynogenetic form consisting of females only. There appears to be a certain regularity in the distribution of unisexual and bisexual populations within the limits of natural distribution. Populations in the eastern part of the area are bisexual with a predominance of females, the percentage of males varying considerably. Equal proportion of sexes has been registered only in a few cases. Evidently the numerical predominance of females in bisexual populations results from the mixing of the unisexual and bisexual forms in these populations. The factors determining the percentage of males in these gynogenetic populations are still unknown.

Further to the west, bisexual populations give way to purely female populations. Males are found in these populations as exceptions and do not comprise more than 1–3 percent. Such populations are common in West Siberia, the Urals and the European part of the U.S.S.R. In the countries of West Europe are found populations of both unisexual and bisexual types, bisexual being rarer.

In mixed populations of C. auratus gibelio, gynogenetic females mate with males of the bisexual form of their own species, and in unisexual populations with males of related species. In pond fisheries, offspring of gynogenetic females are usually obtained with the help of young male carp. Unisexuality of C. auratus gibelio was observed by I.A. Anichenko in 1939–1940 (Golovinskaya and Romashov, 1947); but an explanation of this phenomenon was obtained only after the discovery of natural gynogenesis in this species by Romashov and Golovinskaya who obtained experimental evidence of this type of reproduction by detailed genetic analyses of females of the unisexual form (Krushinsky, 1946; Golovinskaya and Romashov 1947). These were immediately followed by cytological research into the processes of fertilization of the unisexual form of C. auratus gibelio, which confirmed the earlier findings (Golovinskaya, 1954).

Genetic Analysis of the bisexual form of C. Auratus Gibelio.

Investigation of the bisexual form of C. auratus gibelio was carried out with the population of the Volma fisheries (in the Byelorussian Soviet Socialist Republic), which was brought from the Amur in 1948 (Golovinskaya, 1960; Golovinskaya, et al., 1965; Tcherfas 1966b). It was found that specimens of the bisexual form of C. auratus gibelio, when mated with each other, yielded equal numbers of female and male offspring.

The diploid number of chromosomes of the bisexual form was found to be 94 (by the Japanese scientist Makino in 1939). Maturing of bisexual females is accompanied by the usual meiosis. In the last prophase of the first meiotic division all chromosomes are represented by bivalents. The first division is reducing, and takes place shortly before ovulation ends with discharge of the polar body. As with all fishes studied, the ovum of the bisexual C. auratus gibelio is in the metaphase of the second meiotic division after ovulation and contains the haploid number of 47 chromosomes. Shortly after insemination the second (equational) division comes to an end and the second polar body detaches itself. The remaining female telophasic group and the sperm head turn into female and male pronuclei. When preparing for the first division of segmentation both pronuclei draw closer to each other. This process ends with the formation of the diploid metaphasic plate of the first division of segmentation.

Cytological Analysis of Natural Gynogenesis of C. Auratus Gibelio

The exclusion of the male nucleus in the reproduction of the unisexual form of C. auratus gibelio, which was quite clearly proved by experimental crossings, was also confirmed by direct cytological observations of the sperm nucleus in the plasma of the ova. The first cytological analysis of the process of fecundation of the unisexual form of C. auratus gibelio was carried out by K.A. Golovinskaya in 1954. Later these results were confirmed in experiments with females of the unisexual form from many populations (Lieder, 1955, 1959; Statova, 1963). Observations showed that the centrosome brought in by the sperm doubles, and each of the two centrosomes forms a pole of the spindle of the first division of segmentation.

However, during the whole period of preparation for the first division of segmentation the male nuclear apparatus remains as a dense chromatin formation and does not turn into male pronucleus as in usual fecundation. In the first minutes after insemination, the head of the sperm penetrates into the plasma and gradually comes closer to the female pronucleus. After that it can be detected in form in the zone of one of the poles of the spindle of the first division of segmentation, and later in one of the blastomeres. Lieder managed to trace the head of sperm up to the stage of eight blastomeres (Lieder, 1959). The further destiny of the male chromatin remains unknown. Evidently it is absorbed by the plasma of the ovum.

By analysing the maturing process of the gynogenetic females in the population of the Volma fishery it was possible to ascertain the cytological mechanism securing the permanent number of chromosomes of the unisexual forms of C. auratus gibelio (Tcherfas, 1966 a and b). Chromosomes in the somatic and sex cells of the Volma unisexual females were found to be 141, triploid set of the basic number, 47. Triploidy has also been confirmed by indirect methods, by determining ploidy by the number of nucleoli in the nuclei of the epithelial cells of the fin border of one-day larvae and by cytometric data (by the size of cells of some tissues). At present all these indices are widely used for determining ploidy without directly counting the number of chromosomes. Cytometrical investigations have shown that the size of erythrocytes and nuclei of triploid females is on the average 1.4 times greater than of diploid bisexual females.

The process of maturation of unisexual females differs considerably from the usual meiotic process, which is evidently closely connected with triploidy. In the late prophase of the first meiotic division all chromosomes are univalent and no conjugation of chromosomes is observed. In the last period of maturation, which ends with ovulation, it is possible to single out several main stages.

i. Stage 1 - concentration of univalents in a limited area of the cytoplasm of the animal pole and formation of an achromatic, multi-pole figure in the zone of chromosomes. This stage is observed immediately after the nuclear capsule is broken and corresponds to prometaphase 1.
ii. Stage 2 - formation of a three-pole spindle and distribution of chromosomes in three groups at the poles.
iii. Stage 3 - transformation of the three-pole spindle into a bipolar spindle. At this stage, conventionally corresponding to anaphase 1, chromosomes either distribute themselves into two equal groups or in proportion of 1:2, forming haploid and diploid groups.

The above process is abortive and results in no reduction. It ends in uniting all univalents into the triploid metaphasic plate of the only equational division of maturation. It is at this stage that the egg undergoes ovulation. Soon after insemination the division of the polar body takes place. The female triploid complex that remains in the ovum plasma gradually turns into a triploid pronucleus. Further, the process results in forming the triploid plate of the first division of segmentation. Thus the stable number of chromosomes of the unisexual form of C. auratus gibelio of the Volma population is secured by the exclusion of the reduction division. Due to the exclusion of crossing-over and reduction which during the usual process of meiosis cause a genetic recombination, the progeny of each gynogenetic female from the Volma fishery is a genotypically uniform clone. Females from the unisexual population of the Yakot fishery (the Moscow region) have proved to be triploid with ameiotic maturation. It may be presumed that the cytological peculiarities of Volma gynogenetic females are characteristic of other populations of the unisexual form of C. auratus gibelio. Experiments to obtain unisexual C. auratus gibelio by parthenogenesis gave no positive results. Unfertilized eggs when submerged in water began irregular divisions and soon perished. In order to develop normally the eggs of unisexual females should be inseminated. No case of parthenogenesis of these fish has been observed. Suspected cases of parthenogenesis of some species of fish require reappraisal.

A comparison of the cytology of gynogenesis in C. auratus gibelio with data on parthenogenesis of other animals shows that natural gynogenesis of C. auratus gibelio is a typical case of triploid apomixis. It should be noted that polyploidy and, first of all, triploidy is quite a common phenomenon under natural parthenogenesis. As in other animals, triploidy of C. auratus gibelio is accompanied by the loss of ability for usual sexual reproduction (Baranov and Astaurov, 1956; Astaurov, 1965).

The origin of triploidy in gynogenetic females of C. auratus gibelio is still not clear. It can be presumed that initially the species had a diploid gynogenetic form, and triploidy came about later as a result of crossing diploid gynogenetic females which produce unreduced gametes with males of the bisexual form of their own species or related species. This assumption is based on the results of the well-known model experiments on the silkworm (Astaurov, 1955; 1956) and of some cases of natural triploidy (Daravsky, 1958; 1962). The assumption is borne out also by the features of the maturation process of unisexual females of C. auratus gibelio. The distribution pattern of univalents in haploid and diploid groups during the maturation process of the ovum may be regarded as a result of their cytological and genetic heterogeneity. Therefore, it is reasonable to presume that the genotype of gynogenetic females is of hybrid allopolyploid nature. When studying the fecundation process of bisexual females of C. auratus gibelio, some instances of spontaneous diploidization of the female chromosome complex, caused by the return of the second polar body into the plasma of the ovum, were observed. This process resulted in spontaneous emergence of triploids in the progeny of bisexual females. Thus there is a second channel of triploidy of C. auratus gibelio which is not connected with hybridization, namely, autotriploidy.

It is comparatively easy to distinguish females of the unisexual and bisexual forms of C. auratus gibelio. This is of importance in investigations on mixed populations of the species. Table II cites generalised comparative results of some crossings of Volma females of the two forms. Distant crossings (1) and crossings with the use of irradiated sperm (2,3) were most suitable for distinguishing the two forms. In both cases the difference between the females is quite distinct during the process of embryogenesis; and, after hatching, by malformation in the progeny of the females of the bisexual form. Identification of females by the size of erythrocytes is also possible.

Natural Gynogenesis of Mollienesia Formosa

Another case of natural gynogenesis among fishes is that of the live-bearing fish, M. formosa, belonging to Poeciliidae (Hubbs and Hubbs, 1932). In nature this species is represented only by a gynogenetic form. In natural conditions, females of M. formosa live together with two related species M. sphenops and M. latipinna, and propagate with the help of males of these species. In morphological characters, females of M. formosa occupy an intermediate position between M. sphenops and M. latipinna. This led to the belief, in spite of evidence of gynogenesis, that M. formosa was a hybrid form.

Like the gynogenetic females of C. auratus gibelio, females of M. formosa also cannot propagate by parthenogenesis. When crossed with males of other species they give only female progeny of M. formosa. During the entire period of investigations on natural populations of M. formosa, only three males were found. Phenotypically they resembled F_1 males obtained by crossing M. sphenops and M. latipinna with females of M. formosa and masculinized by methyltestosteron. It was assumed that emergence of males may result from three types of development: phenotypical sex transformation of females; disturbances in their maturation; and introgression of paternal chromatin. No detailed investigation of these males was carried out (Hubbs et al., 1959; Haskins et al., 1960). Experimental crossings over a period of 14 years from 1932 to 1946 yielded 8 000 gynogenetic specimens of 20 generations.

Males of 50 different species, subspecies and races of Mollienesia and males of five other species belonging to Cyprinodontidae were used for crossing (Hubbs and Hubbs, 1946). Much information on the nature of gynogenesis of M. formosa was obtained as a result of studies in the laboratory, and observations of natural population of the fish, using the method of transplantation of tissue (Kallman, 1962 a and b; Darnell et al., 1967). As known, compatibility of tissue is possible when the donor and the recipient are of related genetic constitution. In experiments with M. formosa the transplants were fins, spleen and heart. Transplantation was considered successful if no detachment of the transplants was observed during the whole period of the experiment. The degree of incompatibility of tissue, i.e. the degree of genetic relationship of the donor and the recipient, was determined by the time interval between transplantation and detachment.

When investigating laboratory populations, two types of transplantation were carried out: (1) transplantation of tissue between parents and their offspring, i.e. between P and F_1 (la-P@& and F_1; lb-PB& and F_1); and (2) between the progeny of the first and following generations obtained from the initial parent pair. In crossings, males of M. sphenops were used. Transplantations of 1a and 2 were always successful. In transplantations of 1b, detachment of the transplants was observed. The results allowed the following conclusions:

i. Gynogenetic progeny of M. formosa do not inherit any antigens from the father.

ii. Progeny obtained from a female is a clone genetically identical with the mother form and with one another.

The method of transplantation of tissue was used for investigating the clone structure of some natural populations. In these experiments the natural populations of M. formosa were found to include different clones. Kallman (1962b) assumed that different clones in gynogenetic populations may be regarded as the result of the gradual effect of mutation.

Cytological observations of natural gynogenesis of M. formosa show that gynogenetic females of M. formosa are diploid, the number of chromosomes being 45–46 (Haskins et al., 1960) which corresponds with the diploid number determined for related M. sphenops, M. latipinna and M. selifera. The diploidy of female M. formosa is also confirmed by the size of cells and their nuclei in fin regenerates and the amount of DNA in somatic cells of M. formosa, M. sphenops and M. latipinna (Meyer, 1938; Rasch et al., 1965).

The cytological pattern of maturation of ova of female M. formosa has not been studied. However, the results of transplantation experiments, that allowed the clone structure of gynogenetic populations of the species to be determined, show that the maturation process of M. formosa, as well as that of C. auratus gibelio, is of the ameiotic type. Some crossings of females of M. formosa with males of M. sphenops and M. vittata (in these cases genetic markers were used) gave hybrids (Kallman, 1964; Haskins et al., 1960). F_1 and F_2 were constituted by females and males having characters of both parents.

When mature they produced no progeny. It was observed that hybrids accepted transplants from donors belonging to the mother clone, whereas graftings from the father to hybrids were not successful (Kallman, 1964). Kallman therefore observed that all tissue antigens of the hybrids were characteristic of the mother form. Regarding these hybrids as amphimictic diploids and trying to explain the mechanism of inheritance, Kallman came to the wrong conclusion, that females of M. formosa were entirely homozygous. Also, as a logical proceeding from these calculations was another wrong conclusion about restitution of the diploidy of gynogenetic females of M. formosa in the first division of segmentation. His conclusions contradicted the data on clone multiplication of female M. formosa. By cytophotometric analysis, other investigators (Rasch et al., 1965) have shown that these hybrids are triploids, with the diploid set of chromosomes from the mother and the haploid set from the father. Study of spermatogenesis of hybrid males from F_1 has shown that their sterility is connected with the irregular distribution of chromosomes in meiosis. This explains the results of crossing.

Analysis of the Phenomenon of Natural Gynogenesis in Fish

It is necessary to emphasize that triploidy is characteristic of gynogenetic females of C. auratus gibelio and diploidy of M. formosa. This accounts for the difference in the nature of gynogenesis in the two species. Gynogenesis of M. formosa is evidently a result of mass hybridization between two related species (M. sphenops and M. latipinna) and the subsequent effect of natural selection. The origin of gynogenesis of C. auratus gibelio is not so clear. We assume that gynogenesis of C. auratus gibelio is of hybrid nature, and the emergence of triploidy may have been preceded by diploid gynogenetic forms.

In the case of autotriploidy, the possibility of spontaneous emergence of triploids in bisexual populations and their further destiny is of great interest. However, mention should be made of the well-known thesis that gynogenesis under any condition is determined by the presence in the genetic pool of populations of initial species of genes that control individual stages of gynogenetic propagation. Studies of some general features of natural gynogenesis of M. formosa and C. auratus gibelio reveals the evolutionary importance of this method of reproduction of fish.

Change over from the usual bisexual multiplication to parthenogenesis has two main advantages: (i) Possibility of somatization during meiosis, resulting in preservation of the initial valuable genotype in a number of generations; (ii) rise in effectiveness of reproduction. Both these are true in natural gynogenesis of fish. The assumption that genetic advantages of unisexual lines of C. auratus gibelio may be connected with their hybrid constant heterozygous nature and heterosis (which are inherited in a number of generations due to gynogenesis) was made in the first study of natural gynogenesis of this species (Golovinskaya and Romashov 1947). Cytological studies have shown that there is no crossing-over and reduction during the process of maturation of ova of gynogenetic females of C. auratus gibelio; and this supports the above assumption. The above is true for females of M. formosa; clone reproduction of this species has been proved with the help of tissue transplantations.

Increasing effectiveness of reproduction under parthenogenesis is connected firstly with the doubling of the rate of reproduction due to wholly female structure of populations, and secondly with the removal of the loss of time necessary for meeting of two specimens (a male and a female) under usual reproduction. The latter is especially important for animals with low mobility (lizards, for instance) and when the imaginal stage is short (as in the case of some insects).

For gynogenesis fecundation is necessary, and therefore the main factor determining the numerical strength of gynogenetic populations is the number of males able to ensure the reproduction of gynogenetic forms.

Study of natural gynogenesis of different groups of animals has shown that usually reproduction of gynogenetic females depends greatly on males of the bisexual form of the same or related species. This factor limits their population and dispersion, thus depriving them of the main advantages of parthenogenesis. Natural gynogenesis of fish is peculiar in this respect. The results of numerous crossings of unisexual females of C. auratus gibelio and M. formosa have shown their ability to multiply with the participation of males of various species of fish. In experimental conditions, C. auratus gibelio gave progeny when crossed with males of ten different species representing three families. A large variety of species gave gynogenetic progeny of M. formosa It is therefore possible to conclude that in experimental conditions (under artificial fertilization) fish have no limitations in producing gynogenetic progeny by males of different species. Under natural reproduction it is necessary that the ecology of reproduction of gynogenetic females and of the bisexual species living with them should be adequate.

It is also necessary that the numerical strength of the latter should be great. We have evidence of abrupt decrease in the numerical strength of gynogenetic populations of C. auratus gibelio when these conditions were not manifest. In spite of the limiting effect of these factors, because of their ability to propagate with males of many species, natural gynogenesis of fish leads to increase in the numerical strength of the species, which is characteristic of natural parthenogenesis of other animals. The marked predominance of the unisexual form of C. auratus gibelio within the area of its natural distribution may be attributed to the above considerations. Further investigations of the origin of triploidy of C. auratus gibelio and cytological analysis of natural populations connected with them, as well as research programmes of experimental triploidy would help clearer understanding of the phenomenon of natural gynogenesis of this fish.

It is also necessary to study factors regulating the numerical strength of the unisexual and bisexual forms of C. auratus gibelio in mixed populations of this species and analyse the possibility of change over from one manner of reproduction to another. It is necessary to investigate the cytology of gynogenesis of female M. formosa.

No investigations have been carried out to determine the function of sperm under natural gynogenesis of fish or other animals. The assumption that the inactivated nucleus of sperm has no effect on the characteristics of the gynogenetic progeny is based only on morphological analysis. Experiments in transplantation of tissue were a considerable step forward in this respect. For C. auratus gibelio this problem is of great practical importance, as males of different species are used in commercial raising of the unisexual form of this fish. There is no information available on the nature of the biological inactivator causing inactivation of the male nucleus under gynogenesis. These and many other problems need further investigation.

Spontaneous emergence of gynogenetic offspring in natural conditions was observed under distant hybridization. Hybrid gynogenesis is based on the incompatibility of plasma and chromosomes of the crossing species. Chromosomes brought in by sperm get eliminated to some extent in the primary phase of development of hybrid embryos. Most of these hybrid offspring are haploids and mosaics with grave malformations. Their development is usually more regular when they resemble the mother type. Individual viable specimens emerging among hybrid offspring fully repeat the mother form. Their emergence may be accounted for by complete elimination of male chromosomes and diploidization of the female chromosome complex. Hybrid gynogenesis of fish is described in the experiments in distant hybridization carried out by Nikoljukin (1952) and Kryzhanovskii (1947; 1956) and Kryzhanovskii et al., 1953. Hybrid gynogenesis proves that in natural conditions mass hybridization facilitates the detection of specimens with hereditary tendency to gynogenesis.

Artificial Gynogenesis

Two unrelated phenomena form the basis of artificial diploid gynogenesis of fish. These are insemination of the ova with genetically inactivated sperm, and diploidization of the female chromosome complex. Genetic inactivation of the male nucleus can be achieved by different means but ionizing radiation is most often used for this purpose. The quite distinct differential effect of this agent on the nuclear apparatus and cytoplasmatic components of the sex cell allows the sperm to be treated with very high doses of irradiation. When treated with high doses of irradiation, male chromosomes become genetically inactive, but this does not destroy the sperm's ability to fertilize the centrosome functioning normally to ensure the regularity of the process of the first divisions of segmentation. Progeny obtained in such crossings consist not only of unviable haploids but individual gynogenetic diploids, as the result of spontaneous diploidization of the female chromosome complex. This phenomenon has been observed in the natural gynogenesis of C. auratus gibelio as well.

The phenomenon of irradiative diploid gynogenesis is known as the Hertwig effect, named after the German scientist who first described it. The essence of the Hertwig effect is that when the normal ovum is inseminated with irradiated sperm, the affection of the embryo increases only when the dose is increased to a certain limit. When this limit is achieved, the affection decreases noticeably, and further increase in the dose of irradiation (up to the limit under which sperms still preserve their ability to fertilize) does not cause further affection of the embryo. Treating with high doses of irradiation produces malformed, unviable offspring with a very few normal individuals. This, paradoxical as it may seem, can be explained easily. The effect of irradiation on the embryo increases to a stage where the nucleus of the sperm still retains (if partly) its genetic functions and is able to take part in development. When the dose of irradiation is brought to the limit which causes complete genetic inactivity of the male nucleus, haploid and, in individual cases, diploid radiative gynogenesis takes place.

Experiments in Radiative Gynogenesis

Radiative gynogenesis of fish was observed for the first time in experiments with trout (Opperman, 1953) and much later with Misgurnus fossilis L. (Neifach, 1959). Special study of radiative diploid gynogenesis of fish was carried out recently in the Soviet Union. Species under investigation were: M. fossilis L., C. carpio L. and some species of Acipenseridae. Eggs and milt in all the experiments were obtained with the help of hypophysial injections. Sperm subject to irradiation was kept in test-tubes under low temperature (4–8°C). Irradiation of the sperm was carried out by means of usual X-ray units. The inseminated eggs were incubated in petri dishes in the laboratory.

When inseminating the eggs with sperm treated with irradiation doses varying from 100 kr to 600–800 kr, it was observed that complete genetic inactivity of the male nuclear apparatus occurs when the sperm is treated with 100–200 kr irradiation doses. These doses were therefore used in all the experiments. Evidence of genetic inactivity (when treating with 100–200 kr irradiation doses) was obtained through the changes in the curve of waste, general lessening of the affection of embryos and distinct haploid syndrome of the offspring, and in some cases, the emergence of individual normal larvae.

In experiments without any special inactivators, diploid gynogenetic offspring are very rare. On the average, they comprise 0.33 percent of the total fertilized eggs, for M. fossilis, about 2 percent for Acipenseridae, and 3 percent for C. carpio (Golovinskaya et al., 1963; Romashov and Belayeva, 1965). The small number of diploid gynogenetic offspring is evidence of the rarity of spontaneous diploidization of the female chromosome complex. Besides, it was found that individual females of C. carpio, and more especially of M. fossilis, vary greatly in their ability to produce gynogenetic diploids.

Characteristics of Gynogenetic Progeny

Most complete data on the characteristics of gynogenetic progeny were obtained from carp reared up to the maturation period. All the specimens reared turned out to be females. They were normal in their external appearance but slow in growth. Six out of eight gynogenetic females when crossed with male carp gave normal progeny. Gynogenetic progeny of Acipenseridae were few, and it was possible to rear the larvae up to the period when they began active feeding. In this period their viability was lower than in the control experiment. It is possible that the low rate of growth of gynogenetic carp and the low viability of sturgeons are the result of their high homozygosity. Data on the normal fecundity of the gynogenetic progeny have proved to be of great value for selection.

Methods of Increasing the Numerical Strength of Diploid Gynogenetic Progeny

As stated earlier, the level of spontaneous diploidization of the female chromosome complex is very low and diploid gynogenetic offspring are very rare. To increase the frequency of diploidization of the female chromosome complex, the method of temperature shock is used. At present this method is used in experiments in ploidy of animals that represent quite different systematic groups, most often of fish and amphibians. Treating the eggs with low and high temperatures in the period of meiotic divisions results in different disturbances in the meiotic process. Diploidization of the female chromosome complex may be caused by the disintegration of the spindle, due to which none of the chromosome sets can form the polar body, or by the return of the polar body into the plasma of the ovum (Rott, 1965; Betina and Rott, 1966; Romashov and Belayeva, 1965). When fertilized in the usual way, such unreduced ova develop into triploid specimens; and when inseminated with genetically inactivated sperm, diploid gynogenesis takes place.

Investigations have shown that the frequency of cases of induced diploidization of the female chromosome complex greatly depends on the individual characteristics of the female and on the experimental conditions. These most essential conditions are: (1) temperature shock, (2) the stage at which temperature treatment is applied and (3) duration of treatment. The output of gynogenetic diploids as a rule varies greatly, being very high under a most favourable combination of the parameters.

Temperature treatment gave good results in experiments on M. fossilis. As has been mentioned above, the average output of diploid gynogenetic offspring of M. fossilis comprises 0.33 percent. When the eggs inseminated with irradiated sperm were cooled at a temperature of 1–3°C for 10 minutes after insemination (the stage of anaphase 2) the number of diploid gynogenetic offspring was increased up to 14.9 percent. In the most successful experiments, diploid gynogenetic offspring comprised 50–62 percent of the total fertilized eggs. In thermal shock experiments when the eggs were warmed up to 34°C for 8 minutes after insemination (at the stage of early anaphase 2), the investigators managed to obtain about 17 percent of gynogenetic diploids (Romashov and Belayeva, 1965).

At present, methods of temperature shock are being sought by experiments on carp. Mastering these methods is of interest for the study of experimental triploidy of fish, particularly for the study of prospects of using polyploidy in fish farming. Experiments carried out on C. carpio and M. fossilis and some Acipenseridae have proved that it is possible to obtain gynogenetic progeny of fish. Investigations have revealed the cytological mechanics and genetic characteristics under radiative diploid gynogenesis. The aim of future experiments is to work out methods of raising the numerical strength of diploid gynogenetic offspring. It is possible to visualize the prospects of using artificial diploid gynogenesis (Golovinskaya, 1965, 1968). This method holds great prospects for intensifying homozygotization in some line of breeding. Within one generation it is possible to obtain a result that otherwise requires a long period of inbreeding, which, due to late maturation of carp, is very impeded. This method undoubtedly will be useful in selection of heterosis combinations. Such heterosis will evidently be possible in crossings of the topcross type.

The practical importance of artificial diploid gynogenesis is that it makes it possible to obtain entirely female progeny, due probably to the homogametic nature of the female sex. This conjecture is confirmed by the fact that all carp of gynogenetic origin were females. Experiments on carp have shown that artificial diploid gynogenesis can be used as a special method, opening wide prospects for in-depth analyses of the cytological nature of the fish or other organism under investigation. In conclusion it is necessary to note an important cytological difference between natural and artificial gynogenesis of fish. In natural gynogenesis the initial number of chromosomes remains the same due to non-exclusion of the first meiotic division. It leads to reproduction of the maternal genotype in an unlimited succession of generations and increases the heterozygosis of offspring under the influence of mutation. In artificial gynogenesis the initial number of chromosomes remains the same due to the exclusion of the second meiotic division which results in segregation, i.e. the progeny is not genetically identical with the maternal form and is characterised by high homozygosity. Along with the method of irradiative diploid gynogenesis, other mechanisms of artificial gynogenesis modelling natural gynogenesis are of great interest for future investigations.

Diversity in Fish Chromosomes

The chromosomes are responsible for the mechanism of inheritance to express the genetic characteristics in successive generations. In 1902, W. S. Sutton state hypothesis that the chromosome provide the physical basis of heredity. Chromosomes are named from the fact that they stained preferentially with certain dyes and show suitable cellular material prepared for microscopic examination. They appear thread like at certain times and are composed of linear complexes of deoxyribonucleic acid (DNA), genetic material proper and histone proteins which have a supporting or structural role. All living organisms have chromosomes. Prokaryotes (bacteria and virus) are different from eukaryotes (plants and animals). The eukaryotic cell has nucleus that carries chromosome complements. The number of chromosomes per cell is the characteristics of the particular species but in case of fishes these chromosome numbers different within same species also observed. These chromosome numbers made up of maternal and paternal origin of two sets so called as diploid. Single set (egg or sperm) of chromosomes called as haploid. The chromosomes numbers varies in fishes.

The chromosome material is called chromatin which is of two types like euchromatin and heterochromatin. The euchromatin stains lightly and heterochromatin stains darkly. Euchromatin contains genes in a linear array like beds on a string and heterochromatin is genetically inert and acting in maintaining structural integrity of chromosomes and regulation of the gene expression. Heterochromatin made of highly repeated simple sequences of DNA. The part of chromosome at the end called telomere and the constriction called centromere. The telomeres are stable entities essential for maintaining the integrity of chromosome threads.

The centromere controls the movement of chromosomes during cell division. Both the telomere and centromere are located in the heterochromatin. The chromosomes again divided as metacentric, telocentric and acrocentric based on the position of centromere in the chromosome. When centromere located at centre, then it is called as metacentric, at or very near to one end called telocentric and acrocentric having one arm very longer than other. Intermediate cases are described by using the prefix â€˜subâ€™. The fish chromosomes are comparatively small in size and have its own characterisrics. The NF value i. e. 'nombre fundamental' or number of chromosome arms is important because it gives the genetic content of a chromosome complement.

The chromosome staining is achieved by using dyes having affinity for DNA. But the different chromosome banding techniques shows the finer structure of chromosomes.

1. Fluorescence banding
2. Constitutive heterochromatin banding
3. Giemsa banding
4. Nucleolar organisers
5. Late replicating DNA
6. FISH (Fluorescent *In- Situ* Hybridization).

The importance of fish taxonomy is not only with description of new forms, but also with placing each form within taxonomic system that shows itâ€™s relationships to other forms .For more than a century, systematists have sought to organise this diversity by studying aspects of their external and internal morphology which have been especially successful in defining species and in organising these species into genera. These groupings have usually been confirmed when examined with cytogenetically approaches.

In the last few decades works have been focused on the field of cytogenetic investigation of fishes, especially in the area of systematics, mutagenesis and aquaculture. The karyotype is the chromosome complement of an individual or related group of individuals, as defined by chromosome size, morphology and number. Though for all somatic cells of all individuals of species, the number of chromosomes is used as an indicator of classification of species of chromosomes and interrelationships within families. The studies of these characters help to investigate the aquatic structure for the investigate the aquatic structure for the population of each species population in each habitat, so it can determine what species are related to each other in an accurate manner?. This may help to facilitate the hybridization between them in the future to improve the strains.

Cytogenetical studies on fish have been useful to provide information concerning evolutionary and taxonomic studies, as well as for the genetic improvement of commercial fish stocks (Gold, 1979). Several techniques have been devised to obtain mitotic chromosomes in fish, ranging from direct preparations to long-term cell culture (Denton, 1973, Ojima, 1982 and Alvarez et al., 1991, among others). Among these methodologies, in vivo procedures, usually time- and cost-saving, have been the most widespread (Egozcue, 1971, Gold, 1974 and Rivlin et al., 1985). However, to perform direct techniques, it is necessary to carry out a previous colchicine treatment of live animals for about 1 h. Very often, this is a not a feasible method, as in the case of delicate/fragile species after lengthy transportation or species requiring specially suited tanks (e.g., marine species). Consequently, most of the available cellular material can be lost (Ozouf-Costaz and Foresti, 1992 and Maddock and Schwartz, 1996). In eukaryotes, chromosomes consist of a single molecule of DNA [Link to visual proof] associated with:

- Many copies of 5 kinds of histones. Histones are proteins rich in lysine and arginine residues and thus positively-charged. For this reason they bind tightly to the negatively-charged phosphates in DNA.
- A small number of copies of many different kinds of non-histone proteins. Most of these are transcription factors that regulate which parts of the DNA will be transcribed into RNA.

Fishes have been the subject of an increasing number of cytogenetic investigations in the areas of systematics, mutagenesis and aquaculture. From about 20,000 - 40,000 fish species estimated to occur on this globe, the basic karyotypic characteristics [i.e., diploid chromosome number (2n) and number of chromosome arms (NF) are known for not more than 1700 species, which represent only about 9% of the total number of species available. Efforts have been made from time to time to update the karyotype list of available species. *Cyprinus carpio L.* is a teleostean species having a tetraploid origin. Inspite of the fact that carp possesses a large number of very small chromosomes, it has been fairly well studied by the cytogenetic researchers.

In a majority of these studies, a diploid chromosome number has been reported to be 2n = 100 in common carp (Raicu *et al.*, 1972; Denton, 1973; Zan & Song, 1980; Blaxhall, 1983; Labat *et al.*, 1983; Rab *et al.*, 1989; Larka & Rishi, 1991; Anjum & Jankun, 1994; Anjum, 1995). Diploid chromosome number of native carp from Amur river has been divided into eight well-defined groups on the basis of their morphology and a standard karyotype has also been proposed (Rab *et al.*, 1989). C-banding and Silver-NOR staining has also been employed in some cytogenetic studies on common carp (Takai & Ojima, 1982; Ruifang *et al.*, 1985; Sola *et al.*,1986).

Neotropical fishes present a high chromosome diversity showing a wide diploid number variation range, including different levels of ploidies, sex chromosomes, chromosome supernumeraries, and several cases of polymorphisms, related particularly to heterochromatin and NOR sites. Two main general trends of chromosome diversification can be observed among neotropical fishes. First, several fish groups show a chromosome evolution relatively divergent from the point of view of the karyotypic macrostructure. Sister species show conspicuous differences in karyotype structure and most often also in the number of chromosomes.

On the other hand, there are fish groups in which chromosome evolution has been shown to be less divergent, and in this case whole families or even groups of families may share a common karyotype structure and equal number of chromosomes. Several fish groups appear conservative also with respect to the NOR bearing chromosomes. In this case, NOR chromosome location is invariable among species. In contrast, several other groups present wide NOR variability. Sister species may show quite diverse chromosomes bearing nucleolar organising regions. The NOR and heterochromatin relationship is also very diverse among fishes and this may indicate organisational differences involving these chromosome segments. Thus, neotropical fish fauna presents great chromosome variability, verifiable also by NOR studies.

The karyotypes of five species of Scorpaenidae (genera *Scorpenopsis, Dendrochirus* and *Pterois)* from the Indian Ocean are characterised by a diploid set of 48 chromosomes (mainly acrocentric and/or subtelocentric) and by a NOR location on the small arm of a medium-sized pair. All the chromosomes stained uniformly with DAPI, whereas C-banding evidenced a small amount of heterochromatin.

Despite the marked morphological differences among these species, the low degree of diversification of the sets with respect to the ancestral set of teleosts (2n = 48 acrocentric chromosomes) suggests that chromosome morphology has not undergone profound rearrangements during the evolution of these taxa. (chromosome no. structure, definitions of diff. types, diversity, manipulation, importance added)?

In general, cytogenetics studies of crustaceans are relatively few and very difficult to perform because their chromosome numbers are large (Zhang *et al.* 2003, Lee *et al.* 2004), chromosomes are small and theirs shapes are very variable including metacentric, submetacentric, and acrocentric chromosomes (Tan *et al.* 2004).

Marine crustaceans such as the white shrimp *Litopenaeus vannammei* (Dumas & Campos-Ramos 1999) and the Chinese shrimp *Fenneropenaeus chinensis* (Zhang *et al.* 2003), as well as in marine bivalves such as the Japanese oyster *Crassostrea gigas* (Allen *et al.* 1989) were studied to get good results for manipulation of chromosomes.

Consequently, information on the basic genetics of the tropical crayfish *P. llamasi* is necessary not only to reinforce its potential for aquaculture, but also for genetic improvements and conservation. Loricariidae is one of the largest fish families of the world, with about 650 species separated into six subfamilies.

To date, cytogenetic data on only 56 species of this family are available. The lowest diploid number, $2n$=38, was observed in *Ancistrus* n.sp. 1 (Ancistrinae) and the highest diploid number, $2n$=70, was observed in *Rineloricarian* sp. (Loricariinae). The nucleolar organiser regions (NORs) were seen at a terminal position in six species and at an interstitial position in two.

The karyotypic analysis of Loricariinae and Ancistrinae species revealed that these groups exhibit a large diversity of diploid numbers, suggesting the occurrence of intense karyotypic evolution during their evolutionary history.

Although Reis et al. (2003) consider the Loricariidae as the largest family of catfishes in the world, little is known about the karyotypic organisation in this group (Artoni and Ber-tollo, 2001). Meanwhile, some studies provide some cytogenetics information for Loricariinae, Hypop-topomatinae, Hypostominae, Ancistrini and Neoplecostominae.

In spite of the small amount of information compared with the number of the species already described for Loricariidae, the available data demonstrate that this is a group of great interest for cytogenetic studies, due not only to the variation in chromosome number, 2n = 36 in *Rineloricaria latirostris* (Giuliano-Caetano, 1998) to 2n = 96 in *Upsilodus* sp, but also to the occurrence of many chromosomal rearrangements, suggesting a divergent karyotypic evolution.

7

Collocation of Fish Farming and Crop Cultivation

Collocation of Production Period

The ingestive variation of fish is caused by the growth of fish and environmental conditions. Among the environmental conditions, the seasonal change of water temperature is the main factor. The production period, therefore, has to be collocated with this variation of fish ingestion so as to synchronise the daily production of fodder grass with the daily ingestion of fish. The period of the lowest ingestion amount of cultivated fish in Changjiang Drainage is between Feb. and Mar. and between Nov. and Dec. in a year. The peak is between Jun. and Sept. accounting for 50% of the annual total amount.

The production mode of English Ryegrass and Sudan grass nowadays coincides with the demand of fish feeds.

Ryegrass is sowed in Sept. every year and transplanted in Oct. and mowed in Dec. in Changjiang Drainage. It reaches the peak production between Apr. and May. The integrated fish farm can supply itself sufficiently by the production of Ryegrass. Sudan grass is sowed in mid Apr. and mowed for the first time when it grows to 50 cm high just before the withering of rye grass. It reaches the peak production between June and Sept. The daily output is about 100 kg/mu. Based on the experiments the annual production of 2 grasses can reach about 15,000 kg/mu if it's well managed. The other ways of collocation are as follows:

- To intercrop gramineous grasses: Ryegrass, Sudan grass and leguminous fodder grasses.

The unit area output of two families of fodder grass can be increased and the quality can be improved.

- To adopt ensiling method if there is a surplus of ryegrass or to pelletize ryegrass with other materials, so as to be supplied during crop change in June as supplements.
- To adopt sowing and transplanting by stages and to mow the grass in turn.

In Changjiang Drainage the seeds of Sudan grass are sowed by stages in April ahead of the usual time and can be mowed in mid May. In order to maintain the fertility of the land, pond silt must be fully utilised with some additional organic fertilizers apart from planting leguminous grass, and at last, the grasses should be harvested with stubble remaining in the field.

The Ratio of Water Surface and Crop Field

The ratio of water surface and crop fields means the total area of crop fields collocated with one mu of water surface of fish farming. The following factors must be taken into consideration:

a. Source of feeds and fertilizers and their prices.

b. The abundance of land.

c. The yield & the proportion of various spp.

Because the result of grasses is better than the one of grains to feed fish, grass cultivation is more preferable on fish farms. If so, please reference to the following equation:

$$R = \frac{Y_1 \times f_1}{P_1}$$

R = the ratio of one mu of actual water surface and the area of crop fields

Y_1 = net output of herbivorous fish (kg/mu)

f_1 = average food coefficient of green fodder

p_1 = average yield of green fodder (kg/mu).

Stocking Models of Fish in Fish-cum-crop Integration

The main or only source of feeds and fertilizers is the pasture grasses cultivated on the farm. The spp. stocked should be grass carp and Wuchang fish megalobrama spp. as major spp. which occupy 60–70%. Silver carp & Bighead as minor spp. which occupy about 20–30% assorted with a few of omnivorous fish which occupy about 10%.

The omnivorous fish can not only utilise residues and detritus but also can clean the fish pond for herbivorous fish but the ratio should be limited within 10%. In order to get natural food organism ready for Silver carp and Bighead after stocking, it is necessary to mow the grass and make a compost to fertilize the pond water or stock Silver carp and Bighead half a month later after stocking of Grass carp.

If the farm produces more compost, the proportion of omnivorous fish may be increased a little. The use of pond silt in fish-cum-crop integration There are 2 methods to use pond silt in accordance with crop strains and cropping system:

a. Pond silt is directly used as base manure for Ryegrass, etc. The silt should be harrowed and smoothed after it dries a little and then the seeds can be sowed. If it's a transplantation, there's no need to harrow. The seedlings could be interpolated, economizing on much labour and time. If the silt is used as additional manure, it should be topdressed to the roots of the plant. In summer, silt will be bailed with water to the fields. This method is beneficial for the silt to release or diffuse nutrient elements and detritus. It also benefits increasing of dissolved oxygen in the bottom layer of water.
b. Making compost with pond silt and grass : Usually, there's a surplus of pond silt in winter whereas there's a shortage of pond silt in summer. Therefore dig out a pit, 5 meters square with a depth of 1 m near the crop field and then put in pond silt and grass (straw manure and stable matted grass are better) to make compost and cover the pit with mud. After fermentation, the effectiveness of manure is increased. The compost is usually used as base manure, sometimes mixed with water as topdressing.

It's safer to use pond silt as manure. The applying amount is not necessarily limited, with an average rate of 5–15 tons/mu or even a thickness of 35 cm on the soil.

Economic Efficiency of Fish-cum-crop Integration

To Increase Sources of Feeds and Fertilizers: This pattern of integration could provide a considerable amount of feeds for fish farming. The source of feeds is stable. They are much cheaper and of high quality. The cost could be reduced by one third.

To Save Energy : The energy consumption of transportation of purchasing feeds and fertilizers could be reduced.

To Rationally Utilise the Labour: Fish farming is a seasonal work. The input of labour varies greatly. The labour in slack season can be used for crop production. Part time worker can be fully utilised. This offers more jobs and increases the income

To Set up Reasonable Ecological System: Pond silt is used to plant fodder crops which in turn are used to feed fish. Large amount of fish could be produced and pond silt is accumulated again. It's a cycle which can fully utilise the sunlight, land, pond silt and fertile pond water and can improve the ecological condition of a pond. According to some data, grass planting on pond dyke can reduce the erosion of soils by 57%. Fish-cum-crop integration in line with the local conditions means to combine the Chinese traditional aquaculture with modern technology.

"Rotation" of Fish and Grass

Owing to the non-continuity of fish farming production, most of nursery ponds and grow-out ponds have been left fallow for a certain period of time, e.g. a yearling rearing pond has half a year of fallow from Nov. to June next year. During this period, fish ponds can be used to plant green fodder or green manure crops and then to culture fish. This could lead to the full utilisation of fish ponds.

The Method of "Rotation" of Fish and Grass

The "rotation" of fish and grass can be divided into single pond rotation and multi-pond rotation.

Single Pond Rotation

That means the "rotation" of fish and grass in the same pond. This method is usually adopted after transferring fingerlings to other ponds. Drain the pond, expose its bottom to the sun, trim the pond dyke, smooth the bottom, dig a ditch for drainage and then sow seeds of rye grass or barnyard grass or rice, etc. on the bottom and the slope. The seeds of rye grass are broadcasted in Nov. at a rate of about 2 kg/mu while the seeds of barnyard and rice are generally sowed in late Apr. or early May in Jiangsu Province. In order to extend the growing period germinated seeds could be sowed in the 1st 10 days of Apr. at a rate of 5–6 kg/mu. After sowing, get rid of birds and control the water depth and adopt different field management towards different plants. Rye grass is a dry crop which can't endure waterlogging. Barnyard grass seedlings like to grow in shallow water with a depth of about 5–6 cm. During their growth period, they can be mowed as green fodder crops. Ryegrass can be mowed four times, with the yield of 3000–5000 kg/mu; barnyard grass and rice can be mowed once with the yield of about 2000 kg/mu. The last ratooning plant of the green fodder or green manure crops should be submerged for fermentation as manure. But the amount of grass per square meter should be controlled within 4–5 kg. Then add water to about 2 meters high. In this way, it won't cause lack of oxygen in the pond water. 11–15 days later summerlings are stocked. After stocking, keep close watch over the water quality and fill the pond with fresh water or change the pond water in time accordingly.

Barnyard and rice can be mowed and submerged by stages on the basis of their growth and then different summerlings can be stocked by groups.

Multi-pond Rotation

It needs a set of fish ponds, of which, a few are used to grow grass or green manure crops. Here are two ways. Because of short period of fallow of grow-out ponds, a few of fish ponds are drained first and planted with grass as early as possible but stocked with fish later than the other ponds, thus, extending the growing period of the grass; the rest are drained later but stocked with fish first. The fish ponds are rotated to do this each year. If there is a food constraint for fish, one or two fish ponds could be used to grow crops to provide feedstuff for fish in the other fish ponds. The fish ponds are rotated to grow grass each year.

The Effects of "Rotation" of Fish and Grass

The system of rotation of fish and grass can fully utilise the productivity of fish ponds to provide both green fodder crops for fish and green manure crops for fish ponds to propagate large amount of natural food organisms.

According to the practices of Suzhou Municipal Fish Farm, the phytoplankton can reach 3×10^8 ind/1 at the peak time. The growth of summerlings can reach 1.8–2.5 mm/day in the initial month. If the yield of barnyard grass in a pond is 5000kg, 40kg of grass can be transformed into 1 kg of fingerlings averagely. It can save 0.5–0.8 kg of marketable feeds and reduce 10% of the cost.

In this system, crops utilise the nutrients of pond silt, whereas crop nutrients, which return back to the pond after submergence, not only improve the water quality but also increase the soil fertility. Practices prove that the fish yield increases after submergence of crops, e.g. Suzhou Municipal Fish Farm has adopted this method to nurture fingerlings since 1979. In 1982, the average yield increased from 150 kg/mu to 391 kg/mu, increasing by 1.6 times.

Fish Farming Integrated with Aquatic Plant Culture

In a network of rivers like lower reaches of Changjiang River, a fish farm is often near lakes, rivers or water-logging area or in the vicinity of inlet and outlet of irrigation canal. These water bodies are rich in nutrients, esp. effluents from cities and fish farms. In order to utilise these water resources, fish farmers in the southern part of China often culture aquatic plants in these water bodies. The principal aquatic plants are water hyacinth Eichhornia crassipes, water lettuce Pistia stratistes and water peanut or alligator weed Alternanthera philoxeroides, so-called three Ap's; the secondary are duckweed such as (Spirodela polyrhiza) and Wolffia arrhiza and Lemuna minor.

The Ways of Utilisation of Aquatic Macrophytes

The output of aquatic macrophytes is the highest among the green fodder crops for fish. The yield could be about 15 ton/mu, even above 25 ton/mu. The aquatic macrophytes grow too fast and it causes a lot of trouble in the Tropics and sub-Tropics. However, the nutrients of aquatic macrophytes are higher.

Water hyacinth is called "King of Aquatic Plants". So far as protein produced from per unit area is concerned, it is 6–10 times that of soy bean. Aquatic macrophytes are easy to manage with less labour and lower cost. It's said that one labour can manage 50 mu of 3 aquatic plants and can produce 13.1 ton of crude protein in half a year. The cost of one ton of three aquatic plants (including wages) is only about 1 yuan.

The Methods of Utilisation

Three Ap's Directly Serve as Fish Feeds

Three Ap's are processed in different ways to turn into different sized feeds for a variety of fish spp. in different size.

To nurture fry, 3 Ap's should be mashed into grass paste, and the residues of leaves must be filtered out; then the paste can be sprinkled to the whole nursery pond. Table salt, 2—5% of the weight of the plant should be added to the water hyacinth paste to decrease the toxicity of saponin. The equivalent ratio of three Ap's and soy bean in fish farming is 17.5—25 : 1. It needs 5 kg of soy bean to nurture 10,000 ind of summerlings. We can use 87.5—125 kg of 3 Ap's instead. The cost of soy bean is 13–18 times that of three Ap's. Fry will grow fast with higher survivability if three Ap's are used as fish feeds. For example, Zhuang Aquaculture Brigade, Wuxian Country stocked 80,000 fry in 1977. After 16 days of nurturing, the transferred size was 3,33 cm, with 94.5% survivability from the pond fed with water hyacinth paste while the transferred size was 2.93 cm with 92.8% survivability from the pond fed with soybean milk.

To rear fingerlings three Ap's are esp. good for Silver carp and Bighead. 3 Ap's should be mashed into grass paste too but it's not necessary to remove the residues.

To rear adult fish, with herbivorous fish of different spp. in different sizes as major spp. three Ap's are often pulverised by green fodder crops pulveriser and then fed to the fish.

According to the experiments, about 45 kg of three Ap's could be converted into 1 kg of fish. The macrophytes per-mu can be converted into more than 400 kg of fish. If rice and wheat brans are used as feeds, it will need 1600 kg of brans to produce the same amount of fish. The cost of three Ap's is about 10% of that of brans, but three Ap's are less effective if they are not process. Moreover, since the contents of N.P. and K in three Ap's are , the grass paste could serve as manure in fish ponds also.

Three Ap's - Livestock, Poultry-fish

"Three Ap's are also palatable food for various animals in integrated fish farms. They do not need processing or just a simple processing, so the rate of utilisation is very high. 900—1000 kg of three Ap's can rear one piglet to an adult with a body weight of 60–70 kg with a little amount of wheat and rice brans. The excreta of one pig can be converted into more than 40 kg of fish. In Helei Fish Farm, water hyacinth is fed to ducks at a rate of 150 g/duck/day with a little amount of wheat and rice brans. This can save 10,000 Yuan in feeding 22,000 ducks in half a year. The average excrement of each duck is about 52 kg and can be converted into 3 kg of fish. Thus, three Ap's become fish yield through two trophic levels (i.e. feeding levels).

Dyke-oond System

This is a special pattern of fish-cum-crop integration. On some farms, much of the broad dykes is devoted to economic crops: mulberry bushes, sugarcane, fruit trees, tea and rape, etc, or crops, vegetables and grasses. With different crops, this system can be called "mulberry plotfish pond"; "sugarcane dyke-fish pond"; "fruit tree dyke-fish pond" "rice paddy-fish pond", etc. Among these, mulberry plot-fish pond is more popular. Most of dyke-pond systems are distributed in Pearl River Delta and Taihu Lake basin. According to the historical records, mulberry cultivation and fish farming could be traced back to the 5th century B.C., but they did not form a well-linked dyke-pond system until the 16th century. The dyke-pond system in Taihu Lake basin is similar to that in Pearl River Delta, which is related to the common factors of the nature and the society. Take mulberry plot-fish pond system as an example:

Ecological structure of "mulberry plot-fish pond" system This system has a perfect ecological structure. It includes mulberry cultivation, sericulture, silk extraction, and fish farming with silkworm faece, pupae and waste water. Mulberry is the producer; silkworm is the first consumer; fish are the second ones land ecological system. In fish ponds, there are four ways of energy flow:

- Silkworm faeces are directly taken in by fish and part of detritus can be filtered by filter-feeding fish.
- The inorganic nutrients in silkworm faces are utilised by phytoplankton & heterotrophic bacteria and the biomass of phytoplankton and bacteria in turn are eaten by filter-feeding fish directly or indirectly.
- The leftover of input and fish faces are decomposed by hydro microbes, releasing inorganic nutrients; and then, the same process occurs as in (ii).
- At the same time, pond silt which is composed of all kinds of sediments returns to the pond dyke and the new material cycle begins.

The Material Link of "Mulberry Plot-fish Pond"

The link of energy flow in "mulberry plot-fish pond" system is pond silt and feeds.

Pond Silt

Pond silt is the main source of manure for crops on the pond dyke. Each mu of fish ponds can provide pond silt for 1–2 mu of plants on dyke, usually for direct application but making compost in paddy-fish ponds in Taihu Lake basin.

In winter pond mud (much mud, less water) is removed from fish ponds after draining and applied between mulberry bush lines on pond dyke or bailed (with a little water) to pond dykes and spread evenly after dry. After that, winter crops are intercropped between mulberry bush Lines.

In summer and autumn, liquid silt known as "nihua" (less mud, much water) is bailed to the pond dyke after mulberry leaves are picked 1–2 times, which is practised 2–3 times in Pearl River Delta and 5–6 times in Taihu Lake basin respectively every year to increase the soil fertility and when the base thickness of the soil of mulberry plants increase by 5–6 cm, it is beneficial to next crops. The mud on pond dykes are washed into fish ponds by rains; pond silt is supplied to pond dykes with nutrients. The process runs again and again.

Feeds and Fertilizers

Mulberry plots can supply feeds for fish directly or indirectly:

a. To provide feeds directly, including pasture grass, vegetables and mulberry leaves.

   On mulberry dykes, grass and vegetables can be planted after autumn leaves being picked; grasses and vegetables could be harvested at a rate of 3000 kg/mu and they can be converted into about 100 kg of herbivorous fish and 40 kg of other fish.

b. To provide feeds indirectly.

The first one is silkworm wastes which are a mixture of silkworm faeces, worm sloughs and mulberry leave residues. Silkworm wastes are rich in nutrients: organic material 87%; N 2.2–3.5%; P_2O_5 2.0–2.5%; K_20 1.5–2% and several trace elements but with less moisture, so the nutritive contents of silkworm wastes is higher than that in any kind of livestock or poultry manures. They serve as both fertilizers and feeds. One mu of mulberry plot can produce 2400–2500 kg of mulberry leaves, even 5000 kg. 100 kg of mulberry leaves when fed to silkworm can produce about 30–50 kg of silkworm wastes; therefore, 1250 kg of silkworm wastes can be obtained from sericulture with the mulberry leaves from one mu of mulberry plots. The conventional FCC of silkworm wastes in fish ponds of polyculture is 8. Hence, 1250 kg of silkworm wastes can be converted into over 150 kg of fish. The next one is pupae, which are by-products of silk extraction of cocoons.

The protein content of pupae holds 55.8%, crude fat 29.1%. Pupae therefore are good feedstuff for feed-eaters. Its FCC is 1.5–2. In Pearl River Delta, the production of cocoons from one mu of mulberry field is between 160 and 175 kg out of which about 130 kg of pupae can be obtained, and used as fish feeds, producing 90 kg of feed-eater spp. and their assorted spp.

The third one is waste water with large amount of protein detritus and soluable protein, which is obtained when the cocoons are steamed and processed at the factory. According to practices of the masses, 200 kg of waste water can be transformed into 1 kg of fish. More than 2500 kg of waste water can be obtained from the processing of cocoons which are gained from sericulture with mulberry leaves of one mu. Thus, 15 kg of fish can be produced. When the ratio of dyke and pond is 1:1, the total products (except pupae) from dyke can meet the demand of feedstuff for the fish pond with net fish yield of 300 kg.

Integrated Management of Fish-Livestock-Poultry Farming

Fish-Cum-Duck Integration

The model of fish-cum-duck integration is rather common in China. Both land and water surface are the habitats of ducks. Fish-cum-duck integration is to utilise the biological relationship between fish and duck, which is of mutual benefit. It's not only beneficial to the fattening of ducks but also beneficial to the fish farming by providing more organic manures to fish. Thus, the fish yields can be increased. It's apparent that fish-cum-duck integration could result in good economic efficiency of fish farms.

The History and Status Quo of Fish-cum-duck Integration

Duck raising in fish ponds is an ancient practice in Asia and Europe and this has long been practised in China too. The ancient farmers probably didn't know the advantageous effects of duck raising on fish culture. It's more likely that this integrated production pattern was formed incidentally on the basis of swimming propensity of ducks.

In 1934, the German scientist Probst conducted scientific experiments of integrated fish-cum-duck farming for the first time. He raised ducks in Common carp fish ponds and found that 0.9–1.7 kg of fish could be increased by raising one duck. Since the original output of fish in the experimental fish pond was rather low at that time, the increment of fish yields was conspicuous after ducks were raised in the pond; however, the outbreak of World War II kept the results unemployed. After World War II, the food was in shortage, especially animal protein.

The short supply of animal protein urged the development of the intensive commercial fish farming. Hungary, Czechoslovakia, GDR began to conduct integrated management experiments of fish-cum-duck in large scale early or late from 1952 to 1955. So far, integrated management pattern of fish-cum-duck has chiefly been practised in China, Hungary, GDR, Poland and Soviet Union. Fish-cum-duck integration has developed into a fixed model of integrated fish farming.

In recent years, this model either on the scale aspect or on the managerial aspect has been developing very rapidly, especially in the area of a network of rivers like Jiangsu and Zhejing Provinces.

Beneficial Interactions in Fish-cum-duck Polyculture

From biological standpoint, a fish pond is a semi-closed biological unit. In fish ponds, there are many aquatic animals and plants, most of which could be used as natural food organisms of fish; some are predators of fry and fingerlings or detrimental to fish but can be utilised by ducks. If duck raising is conducted in fish ponds, the water surface can also be put in full utilisation.

Fish ponds provide ducks with an excellent environment, which prevents them from infection of any parasitic and other diseases. Ducks like to eat juvenile frog, tadpole, larvae of dragonfly. That means to eradicate predators for fry and fingerlings. Furthermore, the protein content of these natural food organisms of ducks is high. Therefore, duck raising in fish ponds can appropriately reduce the demand for protein in duck feeds. For ducks raised in pens, the digestable protein content in duck feeds must be 16–18%, even higher up to 20% while for ducks raised in fish ponds, the digestable protein content might be reduced to 13–14%. This can save 200–300g available protein each duck, an equivalent to 2–3% of duck feeds. If the water surface is employed to raise ducks, the droppings of ducks directly go into water bodies providing C, N and P elements frequently. Thus, the biomass of natural food organisms in fish ponds can be increased. Here are two merits:

First, the droppings of ducks go directly into the pond water so that the availability of manure will not be lost at all; second, the "manuring" conducted by ducks on the surface is more homogeneous without any heaping of duck droppings. For this reason, duck raising in fish ponds can promote the growth of fish and increase the fish yields; in the meantime, it can naturally eliminate a series of problems which might be caused by pollution of duck excreta in duck pen.

Duck droppings are good organic manures in pond fish culture. The organic substance content in duck droppings are 3–5 times that of human excreta, inorganic substance such as N, 1.5–2.2 times; K_2O, 2.6–3.1 times. It shows that the quality of duck excreta is higher than that of human excreta. However, the quality and quantity of duck excreta are determined by duck species, feeds given, culturing management as well as climatical conditions, etc. In Europe, the stocking rate of ducks is generally 300–500 ducks/ha in summer and each duck produces about 7 kg of droppings during the fattening period of 36 days.

If 500 ducks are raised, the total duck excreta would reach 3000–3500 kg in that period. The moisture content in duck excreta holds 56.6%; organic substance, 26.2%; C,10%; P_2O_5, 1.4%;N,1%; K_2O, 0.62%; Ca, 1.8% and some other elements. The annual excreta of each egg-laying duck is 7.5–10 kg in dry weight, about 70 kg in wet weight. Each egg-laying duck raised in Helei fish farm, Wuxi, produces 40–45 kg of manure in wet weight per annum.

Besides the above mentioned, fish-cum-duck integration is beneficial to fish farming:

1. It can fully utilise the uneaten feeds because 10–20% of feeds are spilt by ducks to the ground. Based on the investigation, the fine feeds lost is 23–30g per duck per day or even higher. These spilt feeds and leftover falling directly into the pond water or flushed into fish ponds are utilised as fine feeds for fish resulting in achieving relevant fish yields. Furthermore, it reduces the wasting of duck feeds.
2. It can promote the recycling of nutrients in pond ecological system. The water surface is usually the duck's habitat. In shallow places, a duck often puts its head into the pond bottom and turns the silt to search for benthos. By virtue of this digging action, the nutritional elements deposited in the pond humus will diffuse. It's beneficial to the material recycling in the pond.
3. Ducks are the volunteer aerators. The swimming, playing and chasing of ducks in a pond more or less play a role in aeration, disturbing the smooth surface of the pond in reality.

Barash et al (1982) carried out an experiment on fish-cum-duck integration, rearing Common carp, Tilapia, Silver carp and Grass carp in an experimental pond of 400 m^2. Ducks were fed with nutrient-balanced feed pellet.

At the beginning of the test, duck manure was applied at a rate of about 85 kg/ha in dry weight. At the end of the test, at a rate of 95 kg/ha in conformity with the growth of ducks. There was no other input of manures and feeds except duck excreta and spilt feeds. Compared with the duck raising in pens, the growth rate, feed efficiency and vitality and cleanness of eiderdowns and skin of the ducks raised in fish ponds are better than that in duck pens. Daily fish yield reached 38.5 kg/ha.

Duck raising in fish ponds compared with duck raising in duck pens has three advantages:

1. The feed efficiency and the body weight of each duck increased. The higher feed efficiency also implies the spilt feeds uneaten by duck could be utilised by fish. Thus, the food conversion rate of fish-cum-duck integration was reduced from 3.84 to 2.64.
2. The survival rate of duck raising in fish ponds increased by 3.3% because fish ponds provided a clean environmental conditions for ducks.
3. Without any other feeds, only the droppings and leftover of ducks made the daily fish output of 36.5 kg/ha. However, someone thinks that if fish and ducks are raised together in the same pond, ducks will eat small fish. Based on many years practices in Helei fish farm and the introduction of duck raising experts, it's easy for ducks to swallow a fish with a body weight below 4 g. Fish with a body weight above 5g have a strong ability to swim and can get away from ducks. In Helei Fish Farm, the recovery rate does not decline when Silver carp summerlings are stocked in fish-cum-duck ponds. Of course, the fry ponds and yearling ponds are not suitable to raise ducks because fish are too small and in high stocking density. Duck raising is generally practised in fingerling ponds or grow-out ponds.

Introduction of Chinese Integrated Fish Farming and its Major Models

The Ratio of Fish-cum-duck Integration

In fish-cum-duck integration, the result of fish farming has much to do with how to raise ducks. There are three types of farming practices:

1. *Raising large groups of ducks in open water:* This is "grazing" type of duck raising. Average number of a group in grazing is about 1000 ducks by one worker. The ducks are generally let loose to go grazing in rivers, lakes and reservoirs during the day, utilising natural food organisms, but are kept in pens at night. This method is advantageous to large water bodies for promoting fish proliferation and can be considered as an integrated management model for large water bodies, such as lakes and reservoirs, etc. As for those fish pond farms, this method of integration can not effectively utilise the duck

manures except the increment of economic benefit from duck raising.

2. *Raising ducks on pond shore:* A relatively large duck shed including a workshop for the administration is constructed in the vicinity of fish ponds with cemented area of dry and wet runs outside. The average stocking rate is about 4 ducks/m^2. The dry and wet runs are being cleaned once a day. During cleaning, the sluice of the wet run is opened to allow organic manure to be flushed into fish ponds through a manure ditch. After this, the sluice is closed and the wet run is filled with fresh water. This method has the advantages for centralised management and adopting mechanisms, but is unable to fully utilise leftover and undigested feedstuff in duck excreta. The effects of leftover and the direct value of feeds more or less are lost. It's also unable to take advantage of the symbiosis of duck and fish.
3. *Raising ducks in fish ponds:* This is the common method of integrated fish-cum-duck farming. The dikes of grow-out or 2-year-old fingerling ponds are partly fenced to form a dry run and part of the water area or a corner of the pond is fenced with used material to form a wet run. The net pen is installed by 40–50 cm above and below the water surface respectively in order to save the net material. In this way, fish can enter the wet run for food while ducks can not escape under the net. In a large pond, a small "island" is constructed at the centre of the pond for installation of demand feeding facilities. The stocking density currently practised in China is higher than that practised in other countries, averaging 4.5 individual/m^2 of pen shed including the dry run and 3—4 ind/m^2 for the wet run.

In the early phase of integrated management of fish-cum-duck, ducks went everywhere in fish ponds to get food; now this pattern has already been improved. The duck raising area has been set up to connect the duck shed, the dry run and the wet run. Whether fish-cum-duck integration succeeds or not primarily depends on technical measures of duck raising.

Both meat ducks and egg laying ducks can be raised in fish ponds. In summer, the 14-day ducklings are accustomed to the life on the water surface. Meat ducks grow fast. The good stock can reach to the marketable size of 2 kg in fish ponds within 48—52 days; the stocks which grow slowly can reach to the marketable size within 55—56 days. Ducks should be on the market as soon as they reach the marketable size or they will lose their feathers, resulting in decrease of food efficiency, body weight and commodity value.

The number of ducks to be raised in this type of polyculture depends on the quantity of duck excreta which in turn are determined by the duck species, quality of feeds given as well as the method of raising them. In raising Beijing ducks, about 7 kg of duck excreta per duck can be obtained during 36-days fattening period.

The egg laying Shaoxin ducks raised in Wuxi produce 42.5—47.5 kg of manure per duck per annum, while the hybrids of Shaoxin and Khaki-Combell ducks produce more than 50 kg per duck. The stocking rate of ducks also depends on the climatical conditions and the stocking ratio and density of the various fish species polycultured in the pond. In Europe, the stocking rate is generally 500ind/ha. As a result, the increment of fish yield will be 90 kg/ha. In tropic and sub-tropic zone, Woynarovich (1980) recommended that the stocking rate should be 2250 ind/ha. In Hongkong, the optimum stocking rate is 2505—3450 ind/ha. In Wuxi, 2000 ind/ha. If the meat ducks are raised in fish ponds, the number should be reduced in view of larger quantity of excreta. In Taihu district, polyculture of 7—8 species of fish is practised in fish ponds, in which the stocking ratio of various species of fish remains unchanged when ducks are raised. If the number of ducks exceeds 3000 ind/ha, then the filter-feeding fish and omnivorous fish could be increased and herbivorous fish could be reduced.

The stacking of organic materials won't take place in this type of integration so long as the stocking rate of ducks is proper and the excreta amount does not exceed the transforming power of a fish pond. Ducks swim loosely in the wet run to search for food, their excreta drop evenly into the wet run and the effects of fertilization of duck droppings expand to the whole pond through winds and waves.

Economic Efficiency

The integrated management of fish-cum-duck farming especially raising ducks on the surface of fish ponds is an economically efficient farming practice. In 1980, Helei Fish Farm conducted a comparative test between pond no. 13 and pond no. 21. Two ponds were adjacent with the same size about one ha. There were pigsties set up on the dyke between the two ponds. Pond no. 13 was stocked with 2207.5 kg of fingerlings and 1900 egg laying ducks without any input of manure; Pond no. 21 was stocked with 2188.5 kg of fingerlings only. The species and size of fish and the feeds given in Pond No. 21 were the same as those in Pond No. 13 but with input of manure at a rate of 20,000 kg/ha/year. The output of Pond No. 13 was 12,234 kg while Pond No. 21 was 10,464 kg.

From the above mentioned comparative test, if the stocking rate of egg laying ducks is 1830—1920 ducks/ha, apart from economizing input of 300000 kg of manure, per ha fish yields can be increased by about 17% over ponds without duck integration. Based on the data and calculation, the average rearing period of ducks is 10 months in a year and about 2.5-5.5 kg of fish might be produced by raising one duck. According to the test in Helei Fish Farm, about 2.5 kg of fish could be produced besides about 200 eggs, (260—300 eggs for the hybrids) by raising one duck in fish pond. Helei Fish Farm raise 20000 ducks every year, providing 850000 kg of duck manure. In 1980, the farm got a net profit of 42000 yuan from duck raising.

There were 48 farmers in the farm. The average income was 881.25 yuan per capita. It's impossible to get an accurate economic analysis on fish-cum-duck integration because the investment, production cost and the yields of duck and fish varies in different countries; even in the same district, fish and duck species, stocking densities, quality and efficiency of feeds, rearing management and climatical conditions differ in thousand ways.

It can be explained by an example: In 1981, Helei Fish Farm raised 22000 ducks. Apart from providing 1,000,000 kg of duck manure and large amount of leftover to fish ponds, it harvested 212,695.9 kg of duck eggs, 6059 kg of duck meat, thus providing the market with 24315 kg of available animal protein which is equivalent to 215560 kg of Grass carp protein. The total annual income was 57740 yuan including 10800 yuan from ducks, 42000 yuan from duck eggs, 4940 yuan from duck excreta for fish farming and it holds 24.62% of total profit of the farm. From the view point of input-output relationship, integrated fish farming with ducks is considered to be the best model in integration of fish, livestocks and poultry. The economic efficiency of fish-cum-pig integration from micro-economic stand point is generally not so high and its profit is low.

In the case of integrated fish farming with poultry, fish-cum-chicken integration lacks symbiotic relationship, which exists in the case of integrated fish-cum-geese farming, but the egg-laying rate of geese and the market demand are rather low. By comparing the protein input and output in integrated fish-cum-duck farming, it's found that the production of 1g of egg protein from Shaoxin ducks requires 5.53 g of feed protein whilst 1g of dairy protein from cows needs 5.55g of feed protein.

They are similar in weight; however, it is relatively easier to raise ducks than cow and the economic efficiency and income of fish-cum-duck far exceeds that of fish-cum-cow, e.g. in 1981, a worker in Helei Fish Farm produced 292.6 kg of protein from cows while his counterpart produced 506.6 kg of protein from ducks which is 73% higher. The net profit per capita in fish-cum-cow farming is 1067.9 yuan and 1427.08 yuan in fish-cum-duck farming, which is 33.6% higher.

Integrated management of fish-cum-duck farming can be further developed to achieve higher economic efficiency by utilising the natural water body to cultivate high-yield aquatic plants as vegetable feeds of ducks and by utilising the wastes of integrated fish farming and from city proper to grow earthworm as animal-based feeds for ducks. At the output end, the products such as fish, ducks and eggs could be further processed before marketing, thus, the economic efficiency and income can be considerably raised either from the angle of energy and nutrient source or economy.

Fish-Cum-Pig Integration

Fish culture combined with pig raising is a traditional integrated fish farming model in China. Now, a few of other countries are also engaged in fish-cum-pig integration. Since 1974, Dr. Buck has conducted experiments on utilisation of pig manure in fish farming. In 1977, he polycultured Cyprinidae in the test pond with Silver carp as major species. The output was 4585 kg/ha.

From economic view point, pig-raising can be divided into three situations: loss, balance and profit. If it combines with duck raising and fish farming, not only the economic efficiency could be increased but also the social efficiency and ecological efficiency could be raised.

The pig food are often leftover and residues from the kitchen, aquatic plants and agricultural products and their wastes. The pig excreta is in turn used as organic manure in fish ponds. It might keep the environment clean. Pork is the indespensable subsidiary food. Pig manure is of high quality by virtue of nutritional completeness. That's why farmers usually combine fish farming with pig raising and it becomes a common pattern of integration in China's rural area.

Methods of Pig Manure Application

There are two types of pig sty in China. One is the simple pig shed constructed on the pond dyke or over the water surface; the other is centralised hog house. Both types have their own merits and demerits. The cost of the simple one is lower and moreover, it's easy to apply the manure to fish ponds. Therefore, this type is more suitable to household or small-scaled farms of fish-cum-pig integration. The excreta of pig can automatically flow or be flushed into the pond. It can save much labour. If the area of a fish pond is less than 8 mu, a pig sty can be set up on the pond dyke and then, pig wastes after flowing into the pond, can diffuse to the whole pond by virtue of winds and waves. If more than 30 pigs are raised in the same spot, the method of flowing by gravity is not suitable because the more the pigs, the more the pig manure.

More often than not the place near the pigsty will be heaped up with manure and the water quality will be partially deteriorated. If pig manure sinks to the bottom too much or flows into the pond too much through the pipe, it will also cause fish surfacing. In order to fully utilise the effectiveness of manure, it's necessary to pay attention to the method of applying. The manure from the centralised hog house is easy to be concentrated to a storage pond or a sedimentation basin. The amount of application can be controlled by various means. Centralised hog house is fit to be built in large-scaled integrated fish farms.

The manure after dilution can be spread along the pond dyke by manual labour or by small boat in small fish ponds. If the fish pond is large, it's better to use boat and mechanic apparatus so as to spread the manure evenly. Here are two methods:

1. To use stern-thruster boat (outboard motor boat) for fertilization Hang the iron cage loaded with manures alongside the boat. The space between grids is about 2.0—2.5 cm. The capacity of the cage differs with the amount of fertilizers needed. The cage loaded with manure must be hanged 10—20 cm below the surface. When the engine starts, the current formed by the thruster will rush against the manure in the cage and as a result, it can be spread evenly.
2. To install a pump in a boat.

The manure should be diluted in a cabin through a funnel and then be spread into the whole pond.

The large quantity of liquid manure in sedimentation basin of largescaled integrated fish farm can be introduced through a pipe or a hose to the pond side and then be sprayed through a nozzle. The average rate is 200 litres/min each nozzle. The nozzle should be fixed

0.5—1.0 m above the surface, in this way, the liquid manure can be fully contacted with air. The nozzle will serve as an aerator if it works to spray water when the D.O.C. in pond water is low.

In fish-cum-pig integration, two points must be especially noticed: one is lack of oxygen in the pond; the other is oligotrophic pond water. That's why the water quality should be monitored at any time. Besides, the production period of pig should match the demand of pig manure in fish farming. Integrated fish farms are chiefly engaged in raising hogs. The advantage is that the production periods of pig and fish farming match well. Hog can be raised for 2 periods a year. The duration of each period lasts for 5 to 6 months. 60% of the total amount of manures applied to fish ponds are given during the first half of a year. The peak of application is in Jan. and Feb. for base manure and in June and July for the additive manure. Usually, no manure is applied after the middle 10 days of Oct.

Therefore, one pig production period is supposed to be from the middle 10 days of Feb. to the middle 10 days of Aug; the other is from the middle 10 days of July to the middle 10 days of Jan. The growth of the first batch of pigs peaks at the latter part of the production period and their excreta as well. It can just meet the fertilizer demand of fish farming. The excreta of pigs from Nov. to Jan. serve as base manure for fish farming next year. This collocation not only can meet the demand of pig manure during the whole period of fish farming but also satisfy the concentrated demand of pig manure in fish farming.

Rate of Pig Raised Per Unit Area of Fish Ponds

The actual rate of pig raised for per unit area of fish pond in integrated fish farms in different areas is from one to five per mu of fish pond. How much excreta of pigs can the given unit area of fish pond accept so as not to deteriorate the water quality of the pond but to maintain sufficient food for fish? It must be worked out through the productive experiments in line with the local conditions. In China, a fish pond is often given various kinds of organic manures such as pig manure, cow dung or other animal manures so the quantity of applying pig manure must be reduced correspondingly.

The capacity of a fish pond to accept animal manures depends upon the techniques of application and the nature of manures. In Hungary, the growth period of fish is 150—180 days. If pig manure is applied by the method of stacking in a corner, the rate of application is 100—134 kg/mu of fish pond. If the method of so-called carbon manuring technique is adopted, that is to say, fresh pig manure mixed with pond water is spread to the whole pond frequently, the rate of application is 20—40 kg/mu/day. It's an equivalent to 1000–1500 kg of condensed liquid manure or 1.2—2.5m^3 of solid wastes of commercial pigsty.

In theory, the maximum capacity per mu to accept manures is 2 to 3 times the above mentioned figures. Why, it involves many facets such as environmental conditions of a fish pond, the quality of pig manure, the managerial techniques of fish ponds and so on.

Fish Species Cultured in Fish-cum-pig Integration

The production efficiency of fish-cum-pig integration appears in the full utilisation of food organisms by fish in pig manured pond. In 1950's, when Hungarian began to culture fish by using pig manure, two problems happened.

Water bloom of Aphanizomenon flosaquae and other blue algae appeared time and again because of non-control of manure application. It causes the disappearance of Daphnia spp.

Common carp can not utilise plankton directly, resulting in waste of primary productivity, The filtering species should be stocked in manured pond, and thus, it will avoid auto shading of phytoplankton and decreasing the products of photosysnthesis.

For this reason, Hungarian have stocked certain kinds of Chinese carps since 1970. Silver carp and Bighead are the best species to control plankton. They have fine filtering organs and can filter and eat phytoplankton in size above 220μ. If the water temperature goes up to 22°C, they can filter and eat more plankton and a great number of detritus and bacterial congregations in a given unit time.

The feed basis formed after fertilization could be utilised more. In mono-pig-manured pond, a few Grass carp may be stocked to control Aphanizomenon flosaqae and submerged vegetation in the pond. In mono-culture of Common carp, 50 kg of pig manure can be converted into 1.25–1.5 kg of Common carp; in polyculture with Common carp as major species, 50 kg of pig manure can be converted into 1.75–2 kg of fish; in polyculture with Silver carp as major species, 50 kg of pig manure can be converted into 3 kg of fish.

The feeding habit of Tilapia is wide in scope and serves as cleaner in pig manured pond. This species fits the pigmanured pond. In Hubei Province, China, Silver carp, Bighead and Tilapia are the main species in pig manured polyculture, assorted with a few Grass carp, Common carp and Xeno cyprinus as minor species.

Economic Efficiency of Fish-cum-pig Integration

The economic efficiency is apparent if the wastes of pig raising can be utilised. In Hunan Province, one mu of fish pond matches with three pigs. In polyculture with Silver carp and Bighead as major species, the output could reach 150–200 kg/mu. In tropic and sub-tropic zone, the temperature is higher and the biological process becomes quicker, more pig manure can be decomposed in the pond and moreover, with proper species in polyculture, the utilisation rate of pig manure will perhaps be much higher.

Fish-cum-pig integration can reduce the production cost. For example, in 1981. Xinan Fish Farm Wuxi gained 3388000 kg of pig manure including 50% flush water. It saved 14229.6 yuan of manure expenditure. The fish yield was 33880 kg. The profit of pig raising itself was not too much, nevertheless, the total revenues of integrated fish farm increased because the excreta of pig took the place of artificial feeds and inorganic fertilizers, which accounted for 58.8% of the cost of fish farming.

However, the simple fish-cum-pig integration becomes less in China. This pattern is usually a part of fish-livestock-agriculture integration system which is a complete man-made ecosystem.

Fish-Cum-Cow Integration

Fish farming using cow manure has long been practiced in China. Since the founding of the People's Republic of China, cow farming has developed rapidly. It promotes fish-cum-cow integration, which is one of the common modellings of integration in the southern part of China. On fish farms cow farming can save much fertilizers, cut down fish feeds and increase the income from milk while fish farming on cow farms can make it realised that "the excreta of a cow can be converted into a thousand jin of fish".

Moreover, cow manure can be disposed at hand resulting in saving much money, labour and energy and also it is good for the environment. If farmers use part time worker to conduct small scale integrated fish farming with cow raising, they can not only earn more money but also can supply both fish and milk to the market.

The Material Basis of Fish-cum-cow Integration

In fish-cum-cow integration, cows are suppliers of the materials for fish farming. They can provide cow manure, leftover and matted grass of cow sheds, etc.

The Biological Basis of Fish Farming Using Cow Manure

Among all the excreta of livestock, the amount of excreta of cows is the greatest and the most stable. If a cow is 460 kg in body weight, it can produce 0.081 kg of manure/kg of body weight per day and the annual amount of faeces can reach 13600 kg and that of urine about 9000 kg.

The nutritive content of cow dung is a little less than that of pig faeces. If 0.024 kg of fresh cow manure is applied to a water body of 1 cubic meter per day, inorganic N and P in fish ponds will be 0.897 mg/L and 0.024 mg/L respectively, which are close to the inorganic N, 0.97—2.06 mg/L (average 1.38±0.42 mg/L) and the inorganic P, 0.018—0.036 mg/L in high-yield fish ponds. The ratio of N/P will be 36.9, which is a little higher than the value of the ratio of N/P in phytoplankton. Apparently it will limit the large increment of phytoplankton in fish ponds. Nevertheless, the average amount of phytoplankton in manured pond can still reach 19.15±6.5 mg/L which is apparently higher than that in control ponds and close to the lower limit of optimum food density (20—100 mg/L) of Silver carp and Bighead. The average biomass of zooplankton will amount to 5.61 mg/L which is much higher than that in unmanured pond and the ratio between the biomass of zooplankton and phytoplankton is 1/3.4.

Both the biomass of zooplankton and the ratio between the biomass of zooplankton and that of phytoplankton come to the level as much as in fertile water fish pond (that is to say, the biomass of zooplankton is between 5 and 30 mg/L and the ratio is between 1/4 and 1/3).

The amount of organic detritus and total bacteria not only surpassed those in unmanured fish pond but also exceeded 22.6% and 8.7% than those in pig-manured pond. It is the increase of natural food organisms, detritus and bacteria in fish ponds that enables the filtering and omnivorous fish to grow faster. According to the experiments of Chinese Freshwater Fisheries Research Centre, the output of Silver carp, Bighead, Common carp and Japanese Crusian carp Carassius carassius in cow-manureed pond is 3.5 times, 2.8 times, 3.3 times, 2.2 times the output in unmanured pond respectively.

The conversion coefficient of cow manure is 3.15 in dry weight or 21 in wet weight at the average manuring rate of 0.17 kg/m^3/week in filtering and omnivorous fish farming and it is 3.3 in dry weight or 26 in wet weight at the same input in Silver carp and Bighead culture. On the basis of the investigation data, about 200 kg of cow urine can be converted into 1 kg of Silver carp or Bighead.

No Risk in Fish Farming Using Cow Manure

Cows are ruminants. It is due to repeated grinding and digestive decomposition catalysed by great amount of micro-organisms in rumina that cow manure is very fine. It can suspend longer in the water. The sinking speed of cow manure particles is 2.6 cm/min and that of pig manure particles 4.3 cm/min. If the same amount is applied, the sediment of pig manure is 33% higher than that of cow manure after 24 hours while the suspended organic detritus below 0.65u in size in cow-manured pond is 39.99 mg/L, which is 153.4% higher than that in pig-manured pond. And it occupies 54.6% in total amount of suspended particles, which is the highest percentage compared with that in pig, duck, or chicken-manured ponds.

The suspensibility not only enables fish to get more feeds but also reduces oxygen consumption caused by manure stacking and avoids forming of harmful gases. The BOD of cow manure is lower than that of other livestock manures because the cow forage has already been decomposed by micro-organisms in cow's body. The BOD of 1 kg of cow manure in five days is 20.6g, 32% lower than that of pig manure which is 30.0g. The same tendency appears when they are in fish ponds.

In one week's consecutive measurement, the lowest dissolved oxygen content in cow-manured pond at 5:30 a.m. is 1.8 mg/L while that in pig-manured pond is 1.0 mg/L. The average dissolved oxygen content is 4 mg/L in cowmanured pond but only 3.3 mg/L in pig-manured pond. By virtue of suspensibility and low oxygen consumption of cow manure, fish farming using cow manure is more safe. Freshwater Fisheries Research Centre conducted an experiment on different livestock manures in fish farming in 1983. It indicated that the survival rate of fish in cow-manured pond is 98%, second to none.

Wasted Food of Cows for Fish Farming

Cows feed chiefly on grass. During the grass growing season of about 7 months, the fresh fodder grass for each adult cow amount to 9000—11000 kg. But the leftover occupies 27.9%, around 3000 kg. However, that period of time happens to be the highest ingestion seasons of herbivorous fish, therefore, the wasted fodder of cow can be utilised as fish feeds. The results differ with different quality of grass. The food conversion coefficiency of terrestrial wild grass is about 40–50. The matted grass in cow shed can only be used as compost of the pond but the leftover of fine fodder for cows can be used as fish feeds.

Management of Fish-cum-cow Integration

Disposition and Proportion

In the integrated fish farm, cow sheds should be built in the vicinity of fish ponds so as to utilise cow manure at hand. The faeces and urine are collected separately, the former transported by convyor or boat or car and spread evenly, the latter pumped into fish pond. If the floor of cow shed is higher than the pond dike, manuring ditch could be dug to collect faeces and urine together, then to flush them into fish ponds. It saves time and labour. But the above methods often cause stacking of manure and uneven fertilization. The best way is to mix up cow dung and urine and to pump them into fish ponds evenly. If the hose pipe has a spray nozzle, the result will be better. If the area of fish ponds is large, the boat dragging a stern manure bucket could be used. Then manure will be evenly spread by the waves caused by sailing.

Proportion here means the area of fish pond matched with one milk cow. It depends on synergism of many factors, such as the amount of cow manure and wasted food; the species ratio and target output of fish; the sediment of pond; the quality and quantity of pond humus; other manures and feeds applied. Supposing fish culture wholly depends upon cow manure and wasted food of cows with a net production target of 250 kg, among which 10% is herbivorous fish, the rest filtering and omnivorous fish, each milk cow can provide manure and feeds for fish in 2 mu of fish pond. If the proportion of feed-eater increases, the method of calculation of fish-cow proportion is shown in the empirical equation related to manure demand and quantity of livestock and poultry in model plan. Period and frequency of manuring stands on the transforming speed of cow manure and season change and the rule of the ebb and flow of natural food organisms in fish pond. In 1983, Freshwater Fisheries Research Centre carried out an experiment on measuring the peak time of N and natural food organisms at different water temperature in a fish pond manured once a week. It showed all nutritive factors had two peaks in a week.

The high peak appeared earlier when the water temperature rose. Therefore, the guidelines to determine period and frequency of manuring are as follows: In late winter and early spring, with sufficient base manure, a few times but in large quantities, that is once every 5–7 days; in spring and autumn, once every three days; in summer, once every day or every two days in small quantities and it's better to manure everyday and make some adjustment according to weather, water colour and fish growth.

Stocking pattern in fish farming by applying the waste in cow raising Cow dung and urine is beneficial to filtering and omnivorous fish culture. Therefore, Silver carp and Bighead are taken as major species assorted with omnivorous fish (Common carp as main minor species), and 15—20 % of herbivorous fish. If it's over that ratio, supplemented feeds must be applied. It's optimal that the output of herbivores occupies around 12% of total net output of the pond.

Economic Benefits of Fish-cum-cow Integration

1. *Both milk and fish harvests:* Each black-and-white milk cow after three births can produce around 5000 kg of milk annually, which contains 155 kg of protein. Except the milk for calves, each cow can not only supply the market with 4500 kg of milk, but also can provide feeds and manures for fish farming, which in turn can produce 500 kg of fish containing 55.8 kg of protein annually. Each cow can provide a total of 210.8 kg of protein that is 3700,000 kcal in calorie value annually. Fish-cum-cow integration can increase protein by 36 % and calorie value by 10.6 % above the unitary cow farming and it can increase protein by 2.7 times and calorie value by 9.5 times over those produced by unitary fish culture.
2. *Increasing revenues, decreasing expenditure:* Fish-cum-cow integration can get more money from fish, milk and calves. The output value of each milk cow is averagely more than 3100 yuan. The excrements and leftover of one cow can be converted into 500 kg of fish without additional feeds and manure. Therefore, it can reduce half of the cost of fish farming.
3. *Alleviating unemployment:* In fish farming, raising four cows can offer one opportunity of employment for people.
4. *Saving energy:* Fish-cum-cow integration can provide feeds and manure for fish farming so that it can save a lot of energy for transportation. However, the investment of fish-cum-cow integration is much greater at one stroke. It needs about 2500 yuan RMB for each milk cow and attached facilities. On the basis of investigation, the profit rate of cow is higher, which is about 30%, that is to say, 100 yuan net profit can be gained from 300 yuan fixed capital. In term of net profits and cost saving in fish farming, the economic return of investment will cover less than three years.

8

Analytical Procedures

Drying Processes

Examples of seafood processes are provided for information only. The National Seafood HACCP Alliance does not endorse or recommend specific seafood processes.

Air dried and pressed mullet roe: Clean roe from blood, gall bags, bits of intestines, and black skin. Wash thoroughly and drain. Roll roe in fine salt, using about 1 pound (454 g) of salt per 10 pounds (4.5 kg) of roe. Remove from salt in 6-12 h and brush well to remove excess salt. Lay roe out to dry in direct sunlight. Turn roe at least every h during the first day and bring roe indoors in the evening. Place boards and weights on the roe during the first night or 2 to compress them slightly. Cure for about 1 week under good drying conditions until the roe is reddish-brown and feels hard. Dip dried roe in melted beeswax. Cool for 15 min, wrap in waxed paper, and store in a cool dry place (Jarvis, 1987).

Bag-shaped dried squid: Remove the head and skin from the body. Turn body inside out and wash to remove ink and other substances. Hang the reversed body on the end of a spit and dry in the sun. After drying for 1/2 d, reverse the body to its normal condition. Shape the body daily until dry (Tanikawa et al, 1985).

Balyk (dried sturgeon meat): Remove the back flesh from the sturgeon. For large fish, cut the back flesh either lengthwise only, or else both lengthwise and crosswise. Place pieces of fish in a tub so they do not touch each other or the sides of the tub. Cover pieces with a thick layer of salt and leave for 9-12 d. Use 2 pounds (907 g) of saltpeter (potassium nitrate) to 1,800 pounds (816.5 kg) of salt to give the fish a reddish colour. If desired, add allspice, cloves, and bay leaves to the brine. Soak the salted sturgeon in freshwater for about 24 h to remove excess salt. Dry 4-6 weeks until a slight mold covers the balyk High quality balyk is soft and tender with a reddish or orange-brown colour, and has an odor something like that of a cucumber. It must be transparent, show no traces of putrefaction, have no bitter taste, and not be too salty (Jarvis, 1987).

Boiled-dried abalone: Remove abalone from shell and trim away viscera. Salt 3-7 d. Wash and boil for 5 min. Dry 7-10 d (Tanikawa et al., 1985).

Boiled-dried sand lance: Wash fish in water to remove scales and impurities. Place fish in baskets and boil in salt water for about 15 min or when the fish float. Use 1-1.2 kg salt to 20 L water. Drain fish and dry on mats 2-3 d (Tanikawa et al., 1985).

Boiled-dried sardines: Use sardines or anchovies about 10 cm long. Wash fish in water to remove scales and impurities. Place fish in baskets and boil in salt water for about 15 min or when the fish float. Use 1-1.2 kg salt to 20 L water. Drain fish and dry on mats for 2-3 d (Tanikawa et al., 1985).

Boiled-dried sea cucumber: Place live sea cucumbers in freshwater for a short time to clean out intestinal tract. Remove intestine from the anus with an eviscerating apparatus. Clean abdominal cavity with a thin brush. Put sea cucumbers in salt water at 3°Bé (Baumé) at 95°C for 1.5-2 h. Deflate bodies that swell during the cooking process. Drain sea cucumbers and remove adhering foam with a spatula-like implement. Straighten body shapes and cool. Roast at 70°C and air-dry for 5 d (Tanikawa et al., 1985).

Boiled-dried scallop: Boil fresh scallops 5-8 min and remove body from shell. Remove mantle and viscera and wash in freshwater to remove sand or pieces of shell. Boil scallops in a salt solution (2.8 kg of salt to 20 L water) 20-30 min. Air-dry 10 d, bringing the scallops inside at night (Tanikawa et al., 1985).

Boiled-dried shark cartilage: Cut cartilage from jaw, fin and head into 7-9 cm lengths. Soak in hot water to remove attached meat. Boil again and air-dry in the sun (Tanikawa et al., 1985).

Boiled-dried shellfish meat: Boil shellfish (oysters, clams, mussels) in seawater to open the shell. Remove shellfish meats and boil in seawater again to increase the firmness of the meats. Air-dry (Tanikawa et al, 1985).

Dried abalone: Remove abalone from the shell. Store meats in about 50° salimeter salt brine for several d to remove mantle fringe and preserve the flesh during drying. Wash and cook for about 30 min in water just below the boiling point. Dry on shallow pans in the sun for 4-5 d, turning at intervals. Cook again for 60 min and dry over a low charcoal fire. Rinse in boiling water and dry in the sun for about 6 weeks (Jarvis, 1987).

Dried clams: Shuck clams and boil in salt water for about 10 min. Spread on trays and air dry 2-3 weeks (Jarvis, 1987).

Dried cod I: Remove head, viscera, and backbone. Dry in the sun until the moisture content is less than 30% (Tanikawa et al., 1985).

Dried cod II: Split cod at the dorsal side and remove viscera. Wash fish with freshwater and remove black membrane from belly cavity. Salt in brine or dry salt. For brining, soak fish in salt solution of 18° Bé that was previously boiled and cooled. Place cod in tank and cover with brine. After 1 d, change brine and press fish down with a weight for 1 d. For dry salting, layer cod with a sprinkling of fine salt between layers. Use about 190-200 kg salt for 1,000 split fish. Salt for 10 d. After salting, wash the fish with a salt solution of 4°Bé and drain. Dry salt the fish again and rinse with freshwater before drying. Air-dry on mats (Tanikawa et al., 1985).

Dried cod fillet: Fillet fish and skin. Salt in brine or dry salt. For brining, soak fish in salt solution of 18°Bé that was previously boiled and cooled. Place cod in tank and cover with brine. After 1 d, change brine and press fish down with a weight for 1 d. For dry salting, layer cod with a sprinkling of fine salt between layers. Use about 190-200 kg salt for 1,000 split fish. Salt for 10 d. After salting, wash the fish with a salt solution of 4°Bé and drain. Dry salt the fish again and rinse with freshwater before drying. Air-dry on mats (Tanikawa et al., 1985).

Dried cod stomach: Remove cod stomach, gullet and gills in 1 piece. Air-dry (Tanikawa et al., 1985).

Dried cuttle fish: Split the head and body and remove the eyes. Wash and dry for 7-8 d on a bamboo blind (Tanikawa et al., 1985).

Dried herring: Remove gills, milt, and viscera. Dry in the sun for 2-3 d. Cut out backbone from the caudal fin to the head and cut off the belly flesh horizontally along the lower line of the backbone. Air-dry for an additional 2-3 weeks (Tanikawa et al., 1985).

Dried herring roe: Soak herring roe in seawater for 4-5 d to remove blood and increase firmness. Wash in a freshwater spray. Drain and air-dry for about a week (Tanikawa et al., 1985).

Dried mullet roe I: Place unbroken roe bags in tubs and sprinkle with salt or soak in brine, using about 5 quarts (4.73 L) of salt per 100 pounds (45.4 kg) of roe. Cure for 10-12 h, drain, and spread on boards in the sun to dry. Take roe in each night to prevent them becoming wet from dew. Dry for about 1 week in fair weather. Dip in 50% beeswax and 50% paraffin and store under refrigeration (Long et al., 1982).

Dried mullet roe II: Place unbroken roe sacs in tubs and sprinkle with salt or soak in brine. Use about 6 kg of salt for 50 kg of roe. Salt for 10-12 h and drain. Place roe on wooden board and cover with a plate and weight. After 1 d, spread roe in the sun to dry. Dry for about 20 d, bringing the roe inside at night (Tanikawa et al., 1985).

Dried octopus: levantine cure: Eviscerate the octopus and wash thoroughly in seawater. Spread octopus out on trays, elevated a few feet above the ground, in the sun. Dry for 10 d to 2 weeks depending on the weather and the size of the octopus (Jarvis, 1987).

Dried octopus: oriental cure: Eviscerate the octopus and wash thoroughly in seawater. Simmer the octopus for about 45 min in water just below the boiling point. Spread octopus out on trays, elevated a few feet above the ground, in the sun. Dry for 10 d to 2 weeks depending on the weather and the size of the octopus. A low charcoal fire can be used in drying (Jarvis, 1987).

Dried pollock: Remove head, viscera, and backbone. Dry in the sun until the moisture content is less than 30% (Tanikawa et al., 1985).

Dried salmon: Break the backbone just back of the head immediately after capture to bleed the fish and to prevent thrashing and bruising of the flesh. Cut off head, leaving the collarbone or nape. Insert a knife at the collarbone and cut along the backbone to within 2-3 inches (5.1-7.6 cm) of the tail. Make a similar cut just under the backbone and break backbone off close to the tail. Scrape out viscera, membranes and other offal, and wash flesh. Make a series of transverse cuts, about 3-4 inches (7.6-10.2 cm) apart, to facilitate drying. Hang fish flesh side out from poles on a drying frame. Put 1 side on each side of the pole. Dry for 10 d to 2 weeks in ordinary weather. Dry longer if drying conditions are not good or the fish are large. Store in a cool dry place (Jarvis, 1987).

Dried-salted jack mackerel, mackerel and saury: Split fish on ventral side and remove viscera. Wash fish in freshwater. Dry-salt overnight and air-dry on mats (Tanikawa et al., 1985).

Dried-salted yellowtail: Split the fish on the ventral side and remove the viscera. Wash with freshwater. Make several lines of half-cuts on the surface of the fish body to allow salt to penetrate easily. Soak the fish in a dilute salt solution to remove blood and other extraneous material. Salt for 4-5 d, using 15-16 kg salt for each 10 big yellowtail. Wash with freshwater and air dry in the sun (Tanikawa et al., 1985).

Dried sardines: Wash small anchovies (6-9 cm long) in freshwater. Spread on mats and air-dry. Turn anchovies several times a day. Move drying fish inside during the night. Dry for several d (Tanikawa et al., 1985).

Dried shark: Remove fins and dress the fish. Wash the dressed fish to remove blood and bits of viscera. Insert 2-3 pieces of cane into the flesh crosswise to hold the fish open. Hang on poles to dry in the air. Dry about 4-7 d (Jarvis, 1987).

Dried shark fins I: Cut fins from sharks. Salt or dust with lime and dry in the sun (Long et al., 1982).

Dried shark fins II: Cut fins off at the joint connecting the fin with the body. Trim away all fleshy parts, leaving only the true fin with its rays. Wash and spread on low bamboo or wickerwork frames to dry in the sun. Turn the fins from time to time. Dry for 2-3 weeks, bringing the fins into a dry shelter at night (Jarvis, 1987).

Dried shark fins III: Cut fins off at the base, avoiding the attachment of the meat as much as possible. Wash in seawater or dilute salt solution with a scrubbing brush. Rinse in freshwater. Bore a hole through the bony part of each fin. Hang fins by a string and air-dry for 2-3 weeks (Tanikawa et al., 1985).

Dried shark fins IV: Remove shark fins and soak in freshwater for 4-5 d. Heat in hot water (90°C [194°F]) for 20-30 min to swell and to remove the epidermis. Cut off cartilage at the base of the fins. Separate the fin rays from the base to the central part by removing the gelatinous substance present between the fin rays. Air-dry (Tanikawa et al., 1985).

Dried shrimp: Wash shrimp and cook in large kettles. Add 10-20 quarts (9.46-19.93 L) of salt to the water per 900 pounds (408.2 kg) of shrimp, depending upon the weather. Use more salt in damp weather than in dry. Put shrimp in water after the water reaches a boil and start the cooking time when the brine again comes to a boil. Cook for 15-45 min, depending upon the size and amount of shrimp and the weather. The shrimp are cooked when there is a clear space between the meat and the shell. Drain the shrimp for 15 min and spread them on drying platforms. Turn the shrimp frequently to promote drying and prevent spoilage. Cover shrimp at night with tarps placed over A-shaped trusses to protect them from rain and dew. Dry for 24-48 h in favourable weather, longer with larger shrimp and high humidity. Remove the shells from dried shrimp mechanically and sift on a coarse wire screen to remove meats (Jarvis, 1987).

Dried shrimp (peeled after drying): Wash shrimp and boil for 30-40 min in salt water (about 360 kg salt per 20 L water). Air-dry the shrimp. Pass dried shrimp through a barrel-like device with revolving short levers to remove shells (Tanikawa et al, 1985).

Dried shrimp (peeled before drying): Remove heads and shells from shrimp. Boil shrimp and air-dry (Tanikawa et al, 1985).

Dried shrimp with shell: Wash shrimp and boil for 30-40 min in salt water (about 360 kg salt per 20 L water). Air-dry the shrimp (Tanikawa et al, 1985).

Dried skates or rays: Lay the fish on its back and make 2 circular cuts down the ventral side. The first slices away the lower wall of the mouth and gill cavity, leaving the wall hanging as a flap. The second cuts away the lower wall of the abdominal cavity, leaving this as a flap. Remove viscera and make a vertical cut from above through the backbone from the head to the base of the tail. Make 1-2 short slashes on each side of the thick base of the tail. Make a series of cuts across the disk of the fish, penetrating to the skin below. Rub sand into each cut and lay the fish in a hole in the beach for about 24 h. Wash the fish, drain, and rub a small amount of coarse sand into the flesh. Hang across pole racks to dry in the open air. Dry for 4-5 d (Jarvis, 1987).

Dried squid: Wash and split the squid. Remove the quill and ink sac. Scrape the inside of the body thoroughly. Spread squid out in the sun to dry, turning at frequent intervals for the first few d. Take squid inside every evening to protect from night fog and dew. Dry for about 10 d (Jarvis, 1987).

Dried squid ("biko-surume"): Cut a hole at the end of the body near the fin. Thrust a bamboo spear through the hole so that the hole remains in the dried product (Tanikawa et al., 1985).

Dried squid ("mizu-surume"): Split body and remove viscera and eyes. Remove most of skin. Stretch body on a kite-shaped frame and dry (Tanikawa et al., 1985).

Dried squid ("niban-surume"): Split the belly side of the mantle from the head to the tail. Remove ink sac and cut the viscera off from the head. Split head and remove the jaws and mouth. Remove quill. Soak in freshwater to whiten the surface of the body. Wash in seawater, or 2-3°Bé brine, to remove mucus and other substances. Rewash in freshwater to remove salt. Hang the squid in the sun with the fin of the body on 1 side and the tentacles on the other. Dry for 3 d if the weather is favourable. Bring the squid inside each evening to protect them from the weather. When the squid are 2/3 dry, shape them to keep the desired appearance (Tanikawa et al., 1985).

Frozen-dried Alaska pollock: Split fish on the ventral side and remove viscera. Soak in freshwater for 2 d to remove blood. Change water 4-5 times/d. Hang fish outdoors to freeze completely. Air-dry for about 70 d (Tanikawa et al., 1985).

Halibut rackling: Remove head and clean fish, leaving the collarbone or nape. Remove viscera, split fish into 2 sides, and remove the backbone. Cut the sides into long narrow strips about 1 inch (2.5 cm) wide, leaving the strips joined together at the collarbone. Wash the pieces thoroughly to remove all traces of blood and drain. Soak the strips in 95° salimeter salt brine for 1-2 h. Hang the fish to dry in a shady place where they will be exposed to as much breeze as possible. Dry for 1-2 weeks Jarvis, 1987).

Moonface-shaped dried squid: Split the belly side of the mantle from the head to the tail. Remove ink sac and cut the viscera off from the head. Split head and remove the jaws and mouth. Remove quill. Soak in freshwater to whiten the surface of the body. Wash in seawater, or 2-3° Bé brine, to remove mucus and other substances. Rewash in freshwater to remove salt. Hang the squid in the sun with the fin of the body on 1 side and the tentacles on the other. Dry for 3 d if the weather is favourable. Bring the squid inside each evening to protect them from the weather. When the squid are 2/3 dry, stretch them side ways into a round shape (Tanikawa et al., 1985).

Salted and air-dried tuna roe I: Wash roe and drain for a few min. Place roe in saturated salt solution and cure for about 12 h. Rinse roe sacs and pierce with a knitting needle to allow moisture to escape. Place roe sacs on a marble slab sprinkled with salt. Scatter additional salt over the roe and place a second marble slab on top. After several h, add additional weight to the top slab. Cure for 2 d, then remove weights, turn over roe sacs, pierce roe sacs again, and sprinkle with fine salt. Replace weights and cure for an additional 4-5 d. Remove weights and rinse roe in strong brine. Hang roe sacs from a line and dry in the shade for several d until they are hard and reddish-brown. Brush with olive oil or coat with beeswax (Jarvis, 1987).

Salted and air-dried tuna roe II: Clean roe and remove oviduct, adipose tissue, and large vein. Force blood out of small veins. Puncture lower end of each sac in several places. Wash thoroughly in seawater, drain for a few min, and cover the roe completely with salt. Cure in the salt for 24-36 h. Rinse roe in seawater and drain. Sprinkle salt on a large board and place roe in rows on the board. Add roe until there are 6-7 layers, sprinkling salt on top of each layer. Place stack in a screw press and apply light pressure. Each day, remove the roe; rinse in seawater, resalt, and return to screw press, applying increased pressure each time. After 9-10 d, wash and scrub with freshwater, and hang in the shade to dry for about 15 d (Jarvis, 1987).

Salted-dried pierced sardines: Wash fish in freshwater and soak in a salt solution of 6-7°Bé for 4-5 h. Pierce the fish through the eyes or from gill slit to mouth with a stick. Dip the fish in freshwater to wash and air-dry for 5-6 d (Tanikawa et al., 1985).

Salted-dried round sardines: Wash fish in seawater. Mix 15-16 kg fish with about 15% salt and cover with water. Salt for 6-8 h, stirring 2-3 times. Pierce each 10 fish from mouth to gill slit with a stick and air-dry. Turn the fish after they are 2/3 dry (Tanikawa et al., 1985).

Salted-dried split sardines: Split fish bodies at the ventral or dorsal sides and remove gills and viscera. Soak fish in a 15-18°Bé salt solution or dry salt over night with 20-30% salt. Wash fish with freshwater, place skin-side down on trays and air-dry. Turn the fish over when they are 2/3 dry (Tanikawa et al., 1985).

Salting and drying cod, cusk, haddock, hake, and pollock: Clean (eviscerate) cod, cusk, haddock, hake, and pollock at sea. Remove heads, split open fish and remove 2/3 of the backbone (that portion from the head to the lower end of the abdominal cavity). Wash fish.

Kench Method

A kench is a regular pile of fish made by laying them on their backs with napes and tails alternating. Spread a considerable quantity of salt over each layer. Turn the top layer of fish with backs up. As the salt extracts the water from the fish it runs to the floor and drains off. Since the fish do not stand in brine it is much more difficult to obtain uniform penetration of salt by the kench method; therefore there is much greater danger of spoilage (souring) by this procedure than by the butt method. Use about 20 pounds (9.1 kg) of salt on each 100 pounds (45.4 kg) of fish.

Water-horsing

Remove fish from butts or kenches and wash with seawater or brine to remove any objectionable slime. Transfer fish to a building or room having a good concrete floor. Kench the fish on frames about 8 inches (20.3 cm) above the floor. Place weights of various kinds on the kenches to press surplus brine out of the fish. Allow the fish to drain and slowly dry in the kenches; the longer they remain on kenches the less time they must remain on the flakes for final drying.

Drying on Flakes

After kenching, place the partially dried fish flesh side up on flakes for further drying. A flake is a rack or lattice bed about 3 feet (0.91 m) wide constructed of triangular strips about 1 inch (2.5 cm) wide (at the base) and nailed about 3 inches (7.6 cm) apart to a substantial framework. Build flakes in the open air about 30 inches (76.2 cm) above the floor. When a rainstorm is imminent, collect the fish in piles and cover with small rectangular boxes with peaked roofs called "flake boxes."

The degree to which the fish are dried depends upon the trade. Fish to be sold in the southern states, must be drier than fish to be marketed locally. Fish for export must be dried as completely as possible. Export fish are dried further in specially constructed heated dryers (Long et al., 1982).

Salting and drying mullet: Split the fish along the back, "mackerel style," so they will lie flat in a single piece, leaving the backbone in. Heads may or may not be removed. Save roe and salt separately. Eviscerate and wash fish to remove all traces of blood from under the backbone and clear away the dark belly cavity skin. If heads are left on, clean out all traces of the gills. Score each fish longitudinally along the backbone and also through the flesh on the topside of the fish. Wash and soak in a light brine solution for about 30 min to remove all traces of blood and slime. Remove from brine and drain for about 15 min.

Use "dairy fine" mined salt and dredge each fish in the salt, rubbing some into the scored cuts on each side. A shallow pan or box about 2 feet (0.61 m) square is convenient for this operation. Pack the salted fish, layer by layer, into barrels or tubs with flesh side up except for the top layer which is packed flesh side down. Sprinkle a little salt on the bottom of the container and over each layer of fish. Place a weight on top of the pack to keep the fish under the surface of the brine that forms. Allow the fish to cure in this brine 36-48 h. Remove fish from brine and drain for15-20 min.

Dry on drying racks made with frames of wood covered with wire mesh and standing on legs 3-4 feet (0.91-1.22 m) high. Dry in the shade under a roof without walls and so located that as much of a current of air as possible will pass over the fish. Oxidation or "rusting" sets in immediately if drying is done under the direct rays of the sun. Lay the salted fish on the racks skin side down, and turn 3-4 times the first day. At night, to prevent spoilage through dampness which causes souring and molding, take fish to a sheltered cover (inside if possible). Drying time usually averages 4 d, but is dependent upon the weather and the size of the fish. The drier the finished product is, the less danger there will be of reddening or rusting. The fish is cured when the surface looks dry and hard and the thumb can be pressed into the thick part of the flesh without leaving an impression (Long et al., 1982).

Skinned dried squid: Wash and split the squid. Remove the quill and ink sac. Scrape the inside of the body thoroughly. Remove about 2/10 of the skin at the end of the body. Dry for about 5 d. Shape the squid before drying is completed (Tanikawa et al., 1985).

Stockfish: Prepare stockfish from cod, haddock, hake, cusk, or coal fish. Split the belly open from the pectoral fins to a little below the anal opening, leaving the isthmus in a solid piece. Cut the head on the ventral side as far as the backbone, following a line just in front of the pectoral girdle. Break the head loose from the body. Remove the viscera and roe, and remove all viscera, blood and other offal. Leave the air bladder intact. Split the fish in 2, except for a short section near the tail, and remove about 2/3 of the backbone. Wash the fish in seawater to remove all traces of blood, liver and kidney. Tie the fish in pairs with a loop of strong twine around the tails. Hand the fish over poles so that the 2 of a pair hang on either side of the drying pole. Arrange poles so that air blows between the rows of fish. Turn the fish every few h. Protect the fish from birds with netting. Fish may also be dried in artificial dryers. Dry the fish until no impression can be made when the thick flesh along the backbone is pressed between thumb and forefinger (Jarvis, 1987).

Trepang (dried sea cucumber) I: Split sea cucumber down the side, scrape clean, rinse, and boiled for 10-30 min, depending on size and variety. Spread cooked sea cucumbers on racks to dry in the sun, turning at frequent intervals during the first part of the drying process. Dry about 20 d in good, clear weather (Jarvis, 1987).

Trepang (dried sea cucumber) II: Boil sea cucumber for about 20 min. Slit open and remove viscera. Place in sun and leave until almost dry. Transfer to smokehouse and smoke for 24 h. Spread smoked sea cucumber on mats in the sun until completely dry (Jarvis, 1987).

Trepang (dried sea cucumber) III: Eviscerate sea cucumber and then boil for 15-20 min in seawater. Wash well with freshwater. Dry in a smokehouse for about 4 d, turning the sea cucumbers frequently during the first day (Jarvis, 1987).

Veziga (dried sturgeon spinal chord): Remove head and tail from fish. Remove spinal chord from the backbone. Pull the spinal chord out at the tail end with a bailing hook and by hand. Wash thoroughly to remove slime and blood. Press out jelly-like material from inside the chord by squeezing the chord between the fingers. Split larger chords lengthwise to remove material. Wash chords in freshwater until clear in colour. Hang chords in current of air until thoroughly dry (Jarvis, 1987).

Fish and Fish Products

The small-scale fisheries of developing countries are vital because they provide a nutritious food which is often cheaper than meat and therefore available to a larger number of people.

In many countries the bulk of fish is sold fresh for local consumption. Processing, where this is done, is either to supply distant markets or to produce a range of products with different flavours and textures.

Fish is an extremely perishable food. For example, most fish become inedible within twelve hours at tropical temperatures. Spoilage begins as soon as the fish dies, and processing should therefore be done quickly to prevent the growth of spoilage bacteria.

Fish is a low acid food and is therefore very susceptible to the growth of food poisoning bacteria. This is another reason why it should be processed quickly. Some methods of preservation cause changes to the flavour and texture of the fish which result in a range of different products. These include:

- Cooking (for example, boiling or frying)
- Lowering the moisture content (by salting, smoking and drying collectively known as curing)
- Lowering the pH (by fermentation)

Lowering the temperature with the use of ice or refrigeration also preserves the fish, but causes no noticeable changes to the texture and flavour.

The Nutritional Significance in the Diet

Fish provides a good source of high quality protein and contains many vitamins and minerals.

It may be classed as either white, oily or shellfish. White fish, such as haddock and seer, contain very little fat (usually less than 1%) whereas oily fish, such as sardines, contain between 10-25%.

The latter, as a result of its high fat content, contain a range of fat-soluble vitamins (A, D, E and K) and essential fatty acids, all of which are vital for the healthy functioning of the body. The table opposite illustrates some of the main nutritional differences between oily and white fish.

Bringing in a Catch of Fish

Average composition of fish

Composition	*White fish e.g. haddock*	*Oily fish e.g. herring*
Energy (KJ)	321	970
Protein (g)	17	17
Fat(g)	07	18
Water (g)	82	64
Calcium (mg)	16	33
Iron (mg)	0.3	08
Vitamin A (m a)	0	45
Thiamine (mg)	0.07	0

Certain processing techniques such as boiling leach the water-soluble vitamins into the surrounding liquid. If this is thrown away, a great deal of nutritional value is lost.

Types of Fish Products

Cooked fish: Cooking provides short-term preservation of fish and it is usually a few days before any deterioration becomes noticeable.

A range of methods are used for cooking fish but the principle of the process remains the same. The flesh of the fish softens, enzymes become inactivated and the process kills many of the bacteria present on the surface of the fish.

Boiling and poaching both involve cooking the fish in hot water whereas frying uses hot oil. The advantage of these techniques is they are very simple and require no more than basic household equipment and are therefore suitable for small-scale production.

Cooked fish products are most usually for immediate consumption and require no sophisticated packaging.

The shelf-life can be extended for a few days by using refrigerated storage and the product should be covered to prevent recontamination.

Cooled/frozen fish: The spoilage of fish is directly related to temperature. The higher the temperature, the faster the spoilage up to around 40°C, above which heat will destroy bacteria and enzymes. Any reduction in the temperature prior to processing will maintain the quality of the fish for longer.

Fish can be kept cool by covering it with clean, damp sacking and placing it in the shade. Although this method is simple and requires no special equipment, the fish still begins to deteriorate within a few hours.

An alternative is to pack the fish with ice. This is an effective method and preserves the fish for a longer period of time. Obtaining ice, however, can be difficult for the following reasons:

- Most ice-making machines are power-operated and therefore require some kind of fuel. Obtaining fuel can often be difficult and the machines may prove expensive to operate.
- A great deal of ice is required and often the cost of the ice is greater than the actual cost of the fish.

Freezing is an alternative method for cooling fish. This technique provides long-term preservation, but it is relatively expensive in terms of equipment and operating costs. In view of this it is not recommended for the majority of small-scale fisheries.

Cured fish products: Curing involves the techniques of drying, dry salting/brining (soaking in salt solution) or smoking. These may be used alone or in various combinations to produce a range of products with a long shelf-life. For example:

- Drying - Smoking - Drying
- Brining - Smoking - Drying
- Salting - Drying
- Salting - Drying - Smoking

Techniques such as these reduce the water content in the flesh of the fish, and thereby prevent the growth of spoilage microorganisms.

Dried Fish

The heat of the sun and movement of air remove moisture which causes the fish to dry. In order to prevent spoilage, the moisture content needs to be reduced to 25 per cent or less. The percentage will depend on the oiliness of the fish and whether it has been salted.

Traditionally, whole small fish or split large fish are spread in the sun on the ground, or on mats, nets, roofs, or on raised racks. Sun-drying does not allow very much control over drying times, and it also exposes the fish to attack by insects or vermin and allows contamination by sand and dirt. Such techniques are totally dependent upon the weather conditions. The ideal is dry weather with low humidity and clear skies.

Drying Fish on Racks

Alternatives to sun-drying involve the use of solar or artificial dryers. There has been a great deal of research on the development of solar dryers as an improved method of drying fish. This has shown that by achieving increased drying temperatures and reduced humidities, solar dryers can increase drying rates and produce a lower moisture content in the final products, with improvements in fish quality compared with the traditional sun-drying techniques.

This was first developed in Bangladesh, but there are now numerous variations in different parts of the world. It is probably one of the most simple designs.

An improved solar dryer, with a separate collector and drying chamber. The chimney is painted black to absorb more heat. This will heat the air inside the chimney, thereby increasing the air flow through the dryer.

An artificially-heated tray dryer. When rain threatens, the trays, which were previously placed in the sun to dry, are assembled on top of each other over a simple heating compartment. A roof and chimney are placed on top, and drying continues by direct heating.

Both solar and artificial dryers try to overcome the difficulties posed by sun-drying during the rainy season. With these dryers it is possible to minimize drying times and to increase the product quality.

It should however, be pointed out that it is only advantageous to use such dryers if there is a market for a higher-quality product or if the fish would otherwise be lost.

Salted fish

Most food poisoning bacteria cannot live in salty conditions and a concentration of 6-10 per cent salt in the fish tissue will prevent their activity.

The product is preserved by salting and will have a longer shelf-life. However, a group of micro-organisms known as 'halophilic bacteria' are salt-loving and will spoil the salted fish even at a concentration of 6-10 per cent. Further removal of the water by drying is needed to inhibit these bacteria.

During salting or brining two processes take place simultaneously:

- water moves from the fish into the solution outside
- salt moves from the solution outside into the flesh of the fish.

Salting requires minimal equipment, but the method used is important. Salt can be applied in many different ways. Traditional methods involve rubbing salt into the flesh of the fish or making alternate layers of fish and salt (recommended levels of salt usage are 30-40 per cent of the prepared weight of the fish).

There is often the problem, however, that the concentration of salt in the flesh is not sufficient to preserve the fish, as it has not been uniformly applied. A better technique is brining. This involves immersing the fish into a pre-prepared solution of salt (36 per cent salt).

The advantage is that the salt concentration can be more easily controlled, and salt penetration is more uniform. Brining is usually used in conjunction with drying.

Ultimately the effectiveness of salting for preservation depends upon:

- uniform salt concentration in the fish flesh
- concentration of salt, and time taken for salting
- whether or not salting is combined with other preservation methods such as drying.

Smoked Fish

The preservative effect of the smoking process is due to drying and the deposition in the fish flesh of the natural chemicals of wood smoke. Smoke from the burning wood contains a number of compounds which inhibit bacteria. Heat from the fire causes drying, and if the temperature is high enough, the flesh becomes cooked.

Both of these factors prevent bacterial growth and enzyme activity which may cause spoilage. Fish can be smoked in a variety of ways, but as a general principle, the longer it is smoked, the longer its shelf-life will be. Smoking can be categorized as:

- *Cold smoking.* In this method, the temperature is not high enough to cook the fish. It is not usually higher than 35°C.
- *Hot smoking.* In this method, the temperature is high enough to cook fish.

Hot smoking is often the preferred method. This is because the process requires less control than cold processing and the shelf-life of the hot-smoked product is longer, because the fish is smoked until dry. Hot smoking does, however, have the disadvantage that it consumes more fuel than the cold-smoking method. Traditionally, the fish would be placed with smouldering grasses or wood. Alternatively, fish may be laid or hung on bamboo racks in the smoke of a fire.

Smoking Fish Traditionally

There are various types of kiln available in different parts of the world, which are used for smoking. Although traditional kilos and ovens have low capital costs, they commonly have an ineffective air-flow system, which results in poor economy of fuelwood and lack of control over temperature and smoke density. Improved smokers include the oil drum smoker and the chorker smoker. As well as improved smokers, there are also improved techniques which involve either pre-salting the fish, so that the moisture content is reduced prior to smoking. Alternatively there are a range of improved kiln and oven designs (for details refer to the equipment catalogue section).

Production of cured fish products: The table below outlines the stages in the production of a range of products:

Packaging of cured fish products: The most important concerns regarding packaging for these products are to prevent moisture pick-up and to prevent recontamination by insects and micro-organisms.

Traditional packaging materials include cane baskets, leaves, and jute bags. Alternatives include flexible packaging such as polythene bags, or wooden and cardboard packs. Indeed, the two may be combined, as in a polythene bag enclosed in an outer cardboard pack.

Fermented fish

Fermentation is a process by which beneficial bacteria are encouraged to grow. These bacteria increase the acidity of the fish and therefore prevent the growth of spoilage and food-poisoning bacteria.

Additionally, salt is used to prevent the action of spoilage bacteria and allow the fish enzymes and the beneficial acid-producing bacteria to soften (break down) the flesh. Fermentation is therefore the controlled action of the desirable micro-organisms in order to alter the flavour or texture of the fish and extend the shelf-life.

The use of fermentation as a low-cost method of fish preservation is commonly practiced all over the world. There are many different types of fermented products and their nature depends largely on the extent of fermentation which has been allowed to take place. They can be categorized as:

- fish which retains its original texture
- pastes
- liquids/sauces.

As with salting, there is little need for equipment other than pans and containing vessels, and the process may easily be carried out on a small scale.

Fish paste (bagoong): This is a product from Eastern Asia. It is made from whole or ground fish, fish roe, or shellfish. It is reddish brown in colour, although this will depend on the raw materials used, and is slightly salty with a cheese-like odour.

Equipment Required

Packaging of fermented fish: There are almost as many traditional methods of packaging fermented fish as there are ways of making it - such as earthenware pots, oil cans, drums and glass bottles. In the past, the latter have been used because of their low cost, but nowadays, cheaper plastic containers tend to replace the traditional types. The most important function of packaging for fermented fish products is that the containers should be air-tight, helping to develop and maintain the airless conditions required for good fermentation and storage. As the major advantage of these products is their low cost, the type of packaging is necessarily restricted. Glass bottles are often used for the better-quality products, but earthenware pots and even plastic bags are used.

Suitability for Small-scale Production

Although traditional processing represents a low-cost option for many small-scale producers, there may be large losses in terms of wasted fish. Improved technologies are usually techniques that require little in the way of expensive equipment, but at the same time increase the quality and the efficiency of the process. Often all that is needed to improve the process and the quality of the final product is the provision of clean water, education and training facilities, simple equipment, or basic materials.

It is important, as with all food processing ventures, to ensure that there is a market for the processed fish. Unfortunately, in areas where fresh fish is a more desirable commodity, small-scale fish processors, with their less-preferred cured products, may face fierce competition from larger-scale processors who have access to refrigeration and transport facilities.

Manual on Fish Canning

Principles of Canning

The technology for preserving foods in cans was developed at the beginning of the nineteenth century when a Frenchman, Nicolas Appert, won a competition initiated by another great character in French history, Napoleon Bonaparte. Napoleon is better remembered for his feats as a conquering General, than he is for providing the stimulus for the development of a food preservation technique that was to mark the start of the canned food industry. Appert won his prize (12 000 francs) for demonstrating that foods which had been heated in air-tight (hermetic) metal cans, did not spoil, even when they were stored without refrigeration.

Once the reliance on the refrigerated and/or frozen food chain had been broken, it was possible to open markets for shelf-stable canned products where no entrepreneur had ventured previously. In the time since Appert's success, the technology of canning has been modified and improved. however.

The principles are as true today as they were when first enunciated. The success of the international fish canning industry rests on the sound application of these principles.

Thermal Destruction of Bacteria

When fish are landed they contain, in their gut and on their skin. Millions of bacteria which. if allowed to grow and multiply will cause a rapid loss of the "as fresh" quality and eventually result in spoilage. During post-harvest handling, in transit to the cannery, the fish inevitably become contaminated with other bacteria; these will further accelerate spoilage unless protective measures (such as icing) are employed. The purpose of canning is to use heat. alone or in combination with other means of preservation, to kill or inactivate all microbial contaminants, irrespective of their source, and to package the product in hermetically sealed containers so that it will be protected from recontamination. While prevention of spoilage underlies all cannery operations, the thermal process also cooks the fish and in some cases leads to bone softening; changes without which canned fishery products would not develop their characteristic sensory properties.

In order to make their products absolutely safe, canned fish manufacturers must be sure that the thermal processes given their products are sufficient to eliminate all pathogenic spoilage micro-organisms. Of these *Clostridium botulinum* is undoubtedly the most notorious, for if able to reproduce inside the sealed container, it can lead to the development of a potentially lethal toxin. Fortunately, outbreaks of botulism from canned fishery products are extremely rare. However, as those familiar with the 1978 and 1982 botulism outbreaks in canned salmon will testify, one mistake in a seasons production has the potential to undermine an entire industry. It is because the costs of failure are so prohibitive that canned fish manufacturers go to great lengths to assure the safety of their products. Safety for the end-user. and commercial success for the canner, can only be relied upon when all aspects of thermal processing are thoroughly understood and adequately controlled.

When bacteria are subjected to moist heat at lethal temperatures (as for instance in a can of fish during retorting), they undergo a logarithmic order of death. Shown in Figure 1 is a plot (known as the survivor curve) for bacterial spores being killed by heat at constant lethal temperature. It can be seen that the time interval required to bring about one decimal reduction (i.e., a 90% reduction) in the number of survivors is constant; this means that the time to reduce the spore population from 10 000 to 1 000 is the same as the time required to reduce the spore population from 1 000 to 100. This time interval is known as the decimal reduction time, or the "D value ". The D value for bacterial spores is independent of initial numbers, however, it is affected by the temperature of the heating medium. The higher the temperature the faster the rate of thermal destruction and the lower the D value - this is why thermal sterilization of canned fishery products relies on pressure cooking at elevated temperatures (>100°C) rather than on cooking in steam or water which is open to the atmosphere. The unit of measurement for D is "minute" (the temperature is also specified, and in fish canning applications it can be assumed to be 121.1°C).

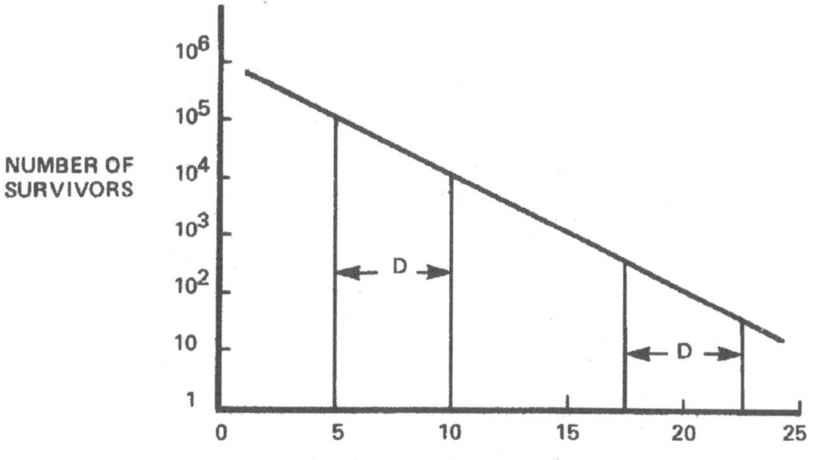

Figure: *Survivor curve for bacterial spores, characterized by a D value of 5 min, subjected to heat at constant lethal temperature*

Another feature of the survivor curve is that it implies that no matter how man decimal reductions in spore numbers are brought about by a thermal process, there will always be some probability of spore survival. In practice, fish canners are satisfied if there is a sufficiently remote probability of pathogenic spore survival for there to be no significant associated public health risk; addition to this they accept, as a commercial risk, the greater probability of there being some non-pathogenic spoilage.

Since it can be seen that not all bacterial spores have the same D values, a thermal process designed to, say, reduce the spore population of one species by a factor of 10 (i.e., 9 decimal reductions or a 9D process) will bring about a different order of destruction for spores of another species. The choice for the fish canner therefore becomes one of selecting the appropriate level of spore survival for each of the contaminating species. Thermophilic spores (those which germinate and outgrow in a temperature range of between 40° and

70°C and have their optimum growth temperatures around 55°C) are more heat resistant, and therefore have higher D values, than spores which have mesophilic optimum growth temperatures (i.e., at 15° to 40°C). This means that raw materials in which there are high levels of thermophilic spores will require more severe thermal processes than will products containing only mesophilic spore formers, if the same degree of thermal destruction is to be achieved for each species.

Thermal Processing Requirements for Canned Fishery Products

From the point of view of preventing microbial deterioration in the finished product. there are two factors which must be considered when a fish canner selects thermal processing conditions. The first is consumer safety from botulism, and the second is the risk of non-pathogenic spoilage which is deemed commercially acceptable.

Table: *Decimal reduction times (D values) for bacterial spores of importance in fish canning*

Organism	***Approximate optimum growth temp. (°C)***	***D value(min) a/***
B. stearothermophilus	55	D121.1 4.0 - 5.0
C thermosaccharolyticum	55	D121.1 3.0 - 4.0
D. nigrificans	55	D121.1 2.0 - 3.0
C. botulinum (types A & B)	37	D121.1 0.1 - 0.23
C.sporogenes (PA 3679)	37	D121.1 0.1 - 1.5
B. coagulans	37	D121.1 0.01 - 0.07
C. botulinum type E	30 - 35 b/	D82.2 0.3 - 3.0

a/ D values quoted are those at the reference temperature of 121.1°C, with the exception of that for *C. botulinum* type E, the spores of which are relatively heat sensitive, being killed at pasteurization temperatures (e.g., 82.2°C)

b/ Although the temperature range for optimum growth of *C. botulinum* type E is 30-35 °C, it has a minimum of 3.3°C which means that it is able to grow at refrigeration temperatures

Safety from botulism caused by underprocessing means that the probability of *C. botulinum* spores surviving the thermal process must be sufficiently remote so as to present no significant health risk to consumers. Experience has shown that a process equivalent to twelve decimal reductions in the population of *C. botulinum* spores is sufficient for safety; this is referred to as a 12D process and assuming an initial spore load of 1 spore/g of product, it can be shown that, for such a process, the corresponding probability of *C. botulinum* spore survival is 10-12, or one in a million million. This implies that for every million million cans given a 12D process, and in which the initial load of *C. botulinum* spores was l/g, there will be only one can containing a surviving spore. Such a low probability of survival is commercially acceptable, as it does not represent a significant health risk. The excellent safety record of the canning industry, with respect to the incidence of botulism through underprocessing, confirms the validity of this judgement. In the United States over the period 1940-82, in which time it is estimated that 30 billion units of low-acid canned food were produced annually (and of these approximately one billion per year were canned seafoods), there have been two outbreaks (involving four cases and two deaths) of human botulism attributable to delivery of inadequate thermal processes in commercially canned food in metal containers. This corresponds to a rate of botulism outbreaks due to failure in the selection or delivery of the thermal process schedule of under 1 in l0 ($0.6/10^{12}$).

Spoilage by non-pathogenic bacteria, although not presenting as serious a problem as botulism will, if repeated, eventually threaten the profitability and commercial viability of a canning operation. It is because of the commercial risks of product failure that canners ought to quantify the maximum tolerable spore survival levels for their canned products. As with the adoption of the 12D minimum process requirement for safety from botulism, experience is the best guide as to what constitutes an acceptable level of non-pathogenic spore survival. For mesophilic spores, other than those of *C. botulinum*, a 5D process is found adequate; while for thermophilic spores, process adequacy is generally assessed in terms of the probability of spore survival which is judged commercially acceptable. In other words what level of thermophilic spoilage can be tolerated bearing in mind the monetary costs of extending processes to eliminate spoilage, the quality costs arising from over-processing and finally the costs of failure in the market place, should surviving thermophilic spores cause spoilage. All things being considered, it is generally found acceptable if thermophilic spore levels are reduced to around 10^{-2} to 10^{-3}/g. There are two reasons why higher risks of spoilage (arising through survival, germination and outgrowth of thermophilic spores) can he tolerated. First, given reasonable storage temperatures (i.e., <35 °C) the survivors will not germinate; and secondly even if spoilage does arise it will not endanger public health.

If a thermal process is sufficient to fulfill the criteria of safety and prevention of non-pathogenic spoilage under normal conditions of transport and storage, the product is said to be "commercially sterile". In relation to canned foods, the FAO/WHO Codex Alimentarius Commission (1983) defines commercial sterility as "... the condition achieved by application of heat, sufficient, alone or in combination with other appropriate treatments, to render the food free from microorganisms capable of growing in the food at normal non-refrigerated conditions at which the food is likely to be held during distribution and storage". Although this definition specifically refers to "non-refrigerated" conditions and thereby excludes those semi-preserved and pasteurized foods in which refrigerated storage is recommended (and in many cases is obligatory in order to prevent growth of the pathogenic psychrophile *C. botulinum* type E -which can grow at temperatures as low as 3.3°C), publications by the Department of Health and Social Security in the United Kingdom and the Standards Association of Australia do not exclude refrigerated foods. According to these less restrictive interpretations, commercial sterility may then also encompass those foods which are intended to be stored at refrigeration temperatures; this implies that commercially sterile canned foods will be free from microorganisms capable of growing at ambient or refrigeration temperatures, whichever is considered normal. Whether the product is intended to be stable under refrigeration or at ambient temperatures, the attainment of commercial sterility is the common objective when manufacturing all canned fishery products. There are, however, circumstances in which a canner will select a process which is more severe than that required for commercial sterility, as for instance occurs when bone softening is required with salmon or mackerel.

The Concept of Thermal Process Severity (Fo Value)

A mathematical equation describing the thermal destruction of bacteria can be derived from the survivor curve shown in Figure 1. If the initial spore load is designated No and the surviving spore load after exposure to heat at constant. temperature is Ns , then the time (t) required to bring about a prescribed reduction in spore numbers can be calculated and is related to the D value of the species in question by the equation,

$$t = D(\log No - \log Ns)$$

From this equation it is apparent that the time required to bring about a reduction of spore levels can be calculated directly, once the spore level before, and the desired spore level after, the heat treatment are specified, and the D value of the spores under consideration is known. For instance, considering the generally recognized minimum process for prevention of botulism through under-processing of canned fishery products preserved by heat alone (which assumes that initial loads are of the order of 1 spore/g, and in line with good manufacturing practice guidelines, final loads shall be no more than 10^{-12}spore/g), the minimum time required to achieve commercial sterility (i.e., a 12D process) can be calculated from,

$t = 0.23(\log 1 - \log 10^{-12})$

$= 0.23 \times$ This means that the minimum thermal process required to provide safety from the survival of *C. botulinum* is equivalent, in sterilizing effect, to 2.8 min at 121.1°C at the slowest heating point (the SHP) of the container. This process is commonly referred to as a "botulinum cook".

Having established the minimum process with respect to product safety, it remains to select a processing time and temperature regime which will reduce the numbers of spore forming contaminants (more heat resistant than those of *C. botulinum*) to an acceptable level. If, for instance, the canner is concerned at the possibility of *C thermosaccharolyticum* spore survival (because it is known that raw materials are contaminated with these spores and it is likely that the product will. be stored at thermophilic growth temperatures) and the No and Ns are 10^2 spore/g and 10^{-2} spore/g, respectively; the time required to achieve commercial sterility can be calculated as before,

$$t = 4.00 (\log 10^2 - \log 10^{-2})$$
$$= 4.00 (2 + 2)$$
$$= 16 \text{ min}$$

Thus, in order to prevent commercial losses through thermophilic spoilage by *C. thermosaccharolyticum* the thermal process must be equivalent, in sterilizing effect, to 16 min at 121.1 °C at the SHP of the container. This approach to calculating the thermal process requirements tends to be an oversimplification for two reasons:

a. in practice it is not reasonable to assume that naturally occurring contaminants will be present only as pure cultures. However, because fish and other raw materials contain a mixed flora, canners assume "worst-case" conditions in order to develop a process which always provides adequate protection from all contaminants. It is customary, therefore, to assume that *C. botulinum* and other heat resistant spore forming bacteria are present: and then to select a thermal process, the severity of which is sufficient to reduce their probability of survival to commercially acceptable levels.

b. The survivor curve (shown in Figure 1) assumes that the temperature of the heat treatment is constant (and in the cases considered, equal to 121.1 °C), whereas during heating in a commercial retort, the SHP of the can experiences a lag in heating and in many cases may never reach retort temperature. Thus the equation that permits calculation of the time required at constant temperature to achieve a desired survivor level (i.e. , Ns) cannot be simply applied to the effects of heating at the SHP of a can. Consequently, the total sterilizing effect at the SHP of a can, which by convention is expressed as time at constant reference temperature, is not the same as the scheduled time for the thermal process (i.e., the time for which a batch retort might be held at operating temperature). To account for the influence on total sterilizing effect of heating lags it is necessary to integrate the lethal effects of all time/temperature combinations at the SHP during a thermal process and express their sum as being equivalent to time at reference temperature. In manufacture of shelf-stable canned fish it is standard practice to express the magnitude of the sterilizing effect of a thermal process in "minutes" at the reference temperature of 121.1 °C. Following this convention, the symbol for the total sterilizing effect of a thermal process is designated as the Fo value; where Fo is defined as being equivalent, in sterilizing capacity, to the cumulative lethal effect of all time/temperature combinations experienced at the SHP of the container during the thermal process. Taking the examples considered above, this means that a botulinum cook must have an Fo value of at least 2.8 min, whereas freedom from thermophilic spoilage by *C thermosaccharolyticum* would necessitate an Fo value of at least 16 min.

Determination of Fo Values

The Fo value of a thermal process can be determined by microbiological or physical means. The former method relies on quantifying the destructive effect of heating on bacterial numbers through their enumeration before and after thermal process; the latter method measures the change in temperature during thermal process at the SHP of the container and relates this to the rate of thermal destruction at a reference temperature. These techniques can be applied to measure the lethal effects of pasteurization processes (in which the target organisms are usually the relatively heat sensitive forms of bacteria, yeasts an moulds) or they may be used to assess the severity of sterilization processes (in which the target organisms are heat resistant spore-forming bacteria). In this text only the physical method of quantifying the lethal effect. of thermal processes will be described.

First, it is necessary to record heat penetration data with thermocouple probes which have been carefully placed to detect changes in product temperature at the thermal centres of the packs. There are many commercial brands of thermocouples available to suit most sizes of fish cans, glass jars and retortable pouches; they can also be constructed with copper/constantan thermocouple wire in which the hot junction is constructed by soldering together the ends of the two wires. The hot junction is coated with a thin laquer layer to insulate the exposed metal surfaces from the product (and thereby prevent surface corrosion which might otherwise interfere with the accuracy of the reading), and then it is carefully positioned at the SHP of the container. Once the thermocouples are in place and the process commenced, the temperature is recorded regularly throughout the heating and cooling phases of the thermal process. The heat penetration data so collected may be treated in a number of ways in order to calculate the Fo value of the process; however, only two of these methods are described in the following sections.

The Improved General Method of Fo Calculation

A plot of temperature versus time is made on specially constructed lethal rate paper in which the temperature (on the vertical axis) is drawn on a semi-logarithmic scale and process time on the horizontal scale; also shown on the vertical axis (but usually, for convenience, on the right-hand side of the paper) is the corresponding lethal rate for the temperature which is on the adjacent left-hand vertical axis. By convention, the rate of thermal destruction (designated L) at product temperature (designated T) for bacteria, or their spores, important in canned fish sterilization is taken to be unity at 121.1 °C; and further, the rate changes by a factor of ten for every 10 °C that the temperature changes. Mathematically this relationship is expressed by the equation,

This means that. the rate of destruction for all temperatures can be related to the rate of destruction at the reference temperature (121.1 °C). Thus the cumulative lethal effects, for all time-temperature combinations experienced at the SHP in a container, can be equated to time of exposure at 121.1 °C.

Once the plot is drawn. the area under the graph is calculated (by counting squares or by using a planimeter) and divided by the area which is represented by 1 min at 121.1 °C .i.e., an Fo value of 1 min. This yields the total sterilizing effect, or the Fo value, of the process. Shown in Figure 2 is an example of a temperature-time plot for a conduction heating pack processed at 121.1 °C. In the worked example, the area under the graph is 70 "units", which when divided by the area corresponding to a Fo of 1 min, i.e. , 4 "units", yields 17.5 min, which is the Fo value for the process being evaluated.

It can be seen that the total sterilizing effect of the process is equivalent to 17.5 min at 121.1 °C, even though the product. temperature never reached 121.1°C, and neither did the retort operate at that temperature. Because it is possible to equate the rates of thermal destruction at any temperature, to the rates of destruction at the reference temperature of 121.1°C, the effects of heating lags can be quantified.

The Trapezoidal Integration and Method

This is a simplified mathematical method in which the time-temperature are used to record the changes in the lethal rates of spore destruction at the SHPs of containers during heating and cooling. If product temperature is recorded at regular time intervals, and assuming that this temperature constant for the period between measurements, the lethal rate applying for time interval can be computed (using equation 1). When the rates (applying over each time interval) are summed and multiplied by the time between measurements, the cumulative Fo value for the entire process can be found without the need graphical representation of the heating and cooling curves. The trapezoidal method also allows simple calculation of the contribution to total process lethality of the heating and cooling components of the process. A worked example in which the product temperature was recorded at 5 min intervals during a process of 60 min at 121.1 °C.

To calculate Fo for the total process: the sum of the L values gives 3.056 which when multiplied by five (the time interval between readings), gives an Fo value of 15.3 min. (Although the theoretical total Fo value for the process is 15.280 min, this can be rounded to 15.3 min as it is unrealistic to quote values beyond the first decimal place.)

To calculate Fo for the heating phase: the sum of the L values at times 25 min and 60 min (i.e., 0 and 0.832) is divided by two and this value (0.416) is added to the sum of the L values from 30 min to 55 min (1.360), so that the total accumulated lethal rate at the time the steam was cut (1.776) can be multiplied by five to give a total Fo value of 8.9 min at steam off. This feature of the trapezoidal method allows for simple calculation of the Fo value during thermal processing, as for instance may be required when the schedule calls for steam to be cut when the Fo reaches an assigned value.

Specification of the Thermal Process Schedule

Once target Fo values for canned fish products are specified. manufacturers must take steps to ensure that all cans receive the correct thermal process and that all factors affecting the rate of heat transfer to the SHP of every can are controlled. It is by these means that microbiological spoilage arising from under-processing can be prevented and the associated health and/or commercial risks avoided. The technique most frequently adopted to control delivery of the thermal process is to draw up a thermal process schedule which specifies those factors which. in any way. could affect delivery of the target Fo value to the SHP of the container. The Codex Alimentarius Commission (1983) destine scheduled process as "the thermal process chosen by the manufacturer for a given product and container size to achieve at least commercial sterility".

Government regulators in many countries adopt similar systems to monitor the scheduled processes of products sold under their jurisdiction. and of these perhaps one of the best known is that implemented by the United States Food and Drug Administration (FDA). In addition to requiring that those processors of acidified and low-acid canned foods sold in the United States register their establishments with the FDA. it is also necessary to file with FDA scheduled processes covering all canned foods which are destined for sale in the United States. Although these requirements will only be relevant for those canners supplying the United States market. the regulations identify several factors which form a useful checklist for canners who are formulating new canned fish scheduled processes. amending existing ones or wishing to review their control procedures. The information which should be specified in the scheduled process is summarized in.

For instance, with some processes the number of retort baskets per retort load will remain constant, whereas with others. it may vary because of delays caused by fluctuations in the supply of fish to the canning line. Under "worst-case conditions" (i.e., with full loads) the steam requirements will be considerably greater than when the retort is only partially full; also. under these conditions steam circulation can be impaired so that the rate of heat transfer to the SHP of the containers is adversely affected. In a case such as this, that steam circulation is influenced by the load size, need be of no consequence. provided the effect is accounted for when calculating the scheduled temperature and duration of the thermal process.

Taking another example, specification of product fill weight may be important when filling solid style tuna or whole abalone into cans which are later to be topped-up with canning liquor; in both instances the convective currents in the brine favour rapid heat transfer to the boundaries of the solid product. there then follows conduction heating during which heat is transferred more slowly to the SHP of the container. However, should fill weight not be controlled. with the result that some cans contain more solid (and therefore less brine. given that the latter is added to a constant headspace). the rate of heat transfer to the SHP of containers will vary, being slower in those packs containing a higher ratio of solids to liquids. The effect of changing the solids to liquids ratio in a pack ought not be underestimated, and alterations should never be adopted without first confirming the adequacy of the process after the proposed change. This point has been demonstrated through trials in which fill weight for solid style tuna packed in 84 x 46.5 mm cans was increased by 10% over the maximum specified. the packs were then processed at 121.1 °C, and in order to achieve a constant target Fo value of 10 min (for the standard and the overweight packs). it was found necessary to increase process time by 16% for the heavier pack.

In this case, failure to compensate for overfilling would not significantly affect public health risks while the target Fo was of the order of 10 min (or more). although there would be an increased probability of survival for those spores more heat resistant than *C botulinum* and. associated with that, an increase in the commercial risk of non-pathogenic spoilage. However, public health risks arising from overfilling can increase for those manufacturers, who, being wary of the reduced yields and or losses in sensory quality caused by processing heat sensitive marine products (e.g. , oysters, mussels and scallops), select target Fo values closer to the minimum for low-acid canned foods (i.e., Fo = 2.8 min).

The rationale behind preparation of the thermal process schedule is to provide a standard format for identifying and specifying all those factors affecting the adequacy of the thermal process. It is important that the scheduled process be developed only by those expert in thermal processing and, only then, when the data upon which recommendations are based are determined in a scientifically sound and acceptable manner. Because of the importance that is attached to correct calculation of thermal processing conditions, it is common to find that in some Countries the regulators overseeing ; canning operations maintain a register of those who are "approved" to establish thermal process schedules.

Once a thermal process schedule has been established it must not be altered without first evaluating the effects of the proposed change on delivery of target Fo values. Also, alterations to product formulation must be evaluated in terms of the possible changes they bring about in the product's heating characteristics. Ideally, specification of the thermal process schedule will be based on data from heat penetration trials with replicate packs, processed under the "worst - case conditions" likely to be encountered in commercial production; however, if this is not possible it is sufficient to refer to those standard texts on canning which recommended process times and conditions for a wide range of canned foods.

In summary therefore, the process schedule provides the specifications which are critical to delivery of an adequate thermal process. The times and temperature of the process schedule is usually contained in the process filing form, an example of which is shown in Figure 3. When completed, the process filing form will also contain additional information which should be specified in the process schedule. It is good practice for the details of the scheduled process to be conveniently located close to the retorts and in a position where it can be seen by the operator.

Application and Control of the Scheduled Process

Once the process schedule is defined, the manufacturer must implement systems to monitor, control and provide records which confirm, after the event, that all stages in production affecting heat transfer to the SHP of the can were within specification. Records provide the means for a continuous assessment of production and an early warning system with which to initiate corrective action if potential problems arise; also they provide valuable and permanent documentary evidence that delivery of the process was in line with details in the process schedule. The value of permanent records becomes apparent at times of product recalls, when the need may arise for the canner to demonstrate that production techniques complied with good manufacturing practice (GMP) guidelines - without this evidence canners risk facing claims of professional negligence should their product become involved in litigation.

Records should be simple to complete, so as not to discourage their use, and easy to interpret. In some cases it may be appropriate to record data on a quality control chart which shows the change in some variable against time. The scales can be chosen to show the change in values about the target value and also include permissible maxima and minima (i.e., tolerances); action levels can be included to alert operators of trends that may cause production to move out of control.

Quality control charts are well suited to continuous operations where monitoring takes place throughout production, they are less frequently used when the function being evaluated is a batch operation. Some recording systems are completed by the operator at specified stages of an operation while others are automated and require only minimal operator input.

No matter what form of records are adapted, their function is to provide retrospective assurance that the thermal process schedule and those related factors which affect heat transfer to the SHP of the container have been regularly monitored and controlled during production.

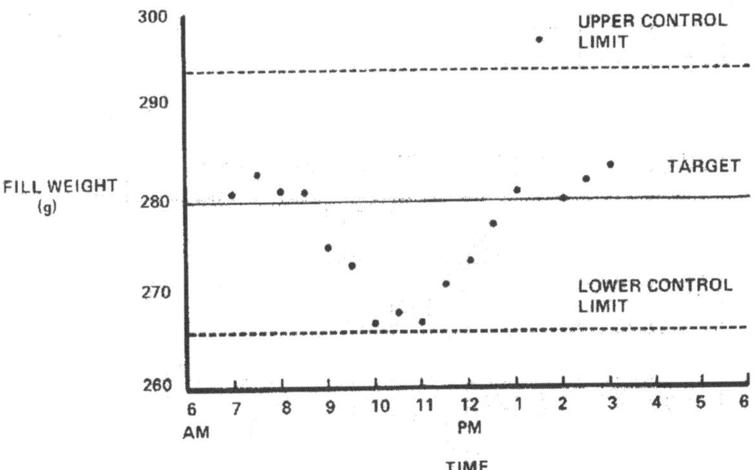

***Figure:** Quality control chart for recording container fill weight*

Packaging Materials for Canned Fishery Products

Containers for thermally processed canned fishery products have several functions in common, whether they are constructed of metal, glass, plastic laminates or composites of plastic and metal laminates. The functions of packaging materials can be summarized as follows:

a. to hermetically seal the product in the container while delivering a thermal process which will render it "commercially sterile";
b. to prevent recontamination of the product after processing. and during subsequent transport and storage; and
c. to provide nutritional benefits and marketing convenience, through presentation of preserved fishery products all year round, often far from the source of supply, and in the majority of cases without the need to rely on refrigerated food chains.

Metal Containers

Tinplate

The most frequently used form of packaging for canned fishery products is tinplate which is fabricated into two and three piece cans of a wide variety of shapes and sizes. Tinplate consists of a base plate of low-carbon mild steel, onto each surface of which is electrolytically deposited a layer of tin. Base plate gauge varies, depending on the size of the cans which are to be manufactured and their intended application; however, it is usually between 0.15 and 0.30 mm thick. Nowadays, for the manufacture of extra light gauge plate, steel sheet is cold rolled twice prior to being tin coated, and in these cases is referred to as double reduced (DR) plate. Tin coating mass varies, according to end use and whether or not lacquers are to be applied; the thickness of the tin coating layers ranges from around 0.4 to 2.5 micron. Plate on which the tin coating mass is the same on each surface is known as equally coated plate; whereas plate with different tin coating masses on each surface is referred to as differentially coated plate.

When specifying tin coating masses it is customary to quote, for each surface, the nominal mass of tin per square metre of plate. Following the standard nomenclature, the designation E05 means that on each surface, there is 2.8 g of tin per square metre of plate; while the designation D10/05 means that the tinplate is differentially, coated and has 5.6 g of tin per square metre of plate on one side, and 2.8 g of tin per square metre on the other surface.

Tin is applied to provide sacrificial protection of the steel base -the tin layer gradually dissolves and passes into the surrounding solution, while the steel layer beneath remains protected. Recently the high cost of tin has made attractive the production of tin-free steel (TFS) in which the conventional tin and tin oxide layers are replaced by chromium and chrome oxide layers. Thus conventional tinplate and TFS consist of multi-layered structures; tinplate comprises an innermost steel layer on top of which there is in sequence. and on each surface of the plate. a tin/iron alloy layer, a free tin layer, a tin oxide layer and an oil lubricant layer, whereas TFS comprises a base steel layer on top of which there is sequentially and on each surface, a chromium layer, a chromium oxide layer and an oil lubricant. Plain TFS cannot be readily soldered, it lacks the corrosion resistance of conventional tinplate (since there is no sacrificial protection of the steel by an overlayer of tin), but it provides an excellent key surface onto which can be applied protective lacquers. Since the introduction of TFS, there has been development of a third system using neither tin nor chromium put nickel as a coating material for the steel base.

In canned fishery products (and with other proteinaceous packs such as meat and corn), it is customary to use sulphur resistant (SR) lacquer systems to prevent the formation of unsightly, yet harmless, blue/black tin and iron sulphides on the plate. Due to the inclusion of white zinc oxide, SR lacquers have a milky appearance. The reason for the inclusion of the zinc is that it reacts with the sulphur compounds, released from the proteins during thermal processing, to form zinc sulphide precipitates which cannot readily be detected against the background of the opaque lacquer.. Another lacquer system finding use for meat and fish packs relies on the physical barrier provided by the inclusion of aluminium pigments in an epoxy-phenolic (epon) lacquer. These lacquers, often referred to as V-enamels, are common in pet food cans.

Aluminium

The dominant position of tin plate as the packaging material of choice for canned seafood products has been challenged with the development of aluminium alloys. Alloys frequently lack the chemical resistance of pure aluminium; however, because they possess greater hardness than that of the pure metal, alloys are well suited to the construction of cans.

The mechanical characteristics, lacking in pure aluminium yet required in the material for food cans, are obtained by the inclusion of small amounts of magnesium and manganese. Depending on the can size and the alloys used, the thickness of the aluminium in fish cans normally ranges from 0.21-0.25 mm. Care must be exercised when manufacturing "easy-open" ends to control the depth of the scores so as to avoid "cut through"; practically, this restricts the lower limit for the, thickness of plate which cap be used in ends.

Aluminium alloys are widely used for the manufacture of dingley, club, hansa, and a variety of conical and straight sided round cans. Some of the important factors which account for the increasing popularity of aluminium for the construction of fish cans are summarized:

- ease of fabrication. Many fish canners manufacture their own can bodies from pre-coated coil stock. thus saving the costs of transporting the bulky empty containers from the can manufacturing plant;
- attractive appearance;
- good corrosion resistance. Although generally more resistant to external atmospheric corrosion, product induced internal corrosion and sulphur staining, than unlacquered tinplate cans, aluminium cans are coated internally with an epon or polyester lacquer and externally with polyesters and polyvinyl fluoride coatings;
- ease of opening tear-off ("easy-open") ends;
- light weight;
- recyclability (however, this characteristic is of greatest significance with carbonated beverage and beer cans);
- elimination of side seams with drawn cans. (This desirable feature is also available with drawn tinplate cans).

Because of their relatively large surface area and flexibility, many aluminium cans ends (e.g. , those on club and dingley cans) are prone to distortion during retorting and early in the cooling cycle (i.e. , when the pressure in the cans is greatest). In some cases this will cause peaking, and it is in order to avoid this that these cans are commonly processed in counterbalanced retorts operating with an overpressure.

Can Construction

Metallic cans are available in a multitude of shapes and sizes to suit all types of canned fishery products. Three-piece cans are manufactured from a rectangular piece of tinplate (known as a body blank) which is formed ifito a cylindrical shape and then joined along a vertical seam by either soldering or welding; to this section are added two ends, one by the can maker and the other. after filling by the canner –the former is referred to as the can maker's end (CME) artd the latter the canner's end (CE).

The seam joining the can end and the b()d.y is known as the double seam and it is the formation of this seal which is critical if the col1tainer is to function correctly. Errors in "double seaming" can lead tb lbS$ of the hermetic seal and the possibility of post-process contamination, giving rise to canned food spoilage. Diagrams illustrating the sequence of rolling double seams, critical doub.1e seam morphology and criteria fot: assessing double seams are presented in section 2.1.4. Experience has shown that the majority of problems arising through faulty double seam formation are associated with errors in application of the canner's ends.

This is attributable to the greater difficulty in applying can ends under commercial filling operations, when compared with completing the same operation in the can making plant.

Two-piece cans for fishery products are made by the draw and re-draw (DRD) process using aluminium or tinplate. While it is possible to have two-piece cans using both tinplate and aluminium (e.g., a tinplate body with an aluminium end), they have the disadvantage that bi-metallic corrosion may occur if the two exposed surfaces come into contact. DRD cans are made from circular blanks of pre-lacquered plate which are first drawn into shallow cups and then re-drawn, once or twice depending on the cans final dimensions, causing an elongation of the wall and a simultaneous reduction of diameter. One great benefit of two-piece cans is that they have no side seam, and only one double seam, thus reducing the risks of leakage arising from imperfect seam formulation.

Aluminium easy-open two-piece cans enjoy great popularity for tanned sardines where the convenience of the tear:-off end is well suited to the dimensions of the product; however the functional benefits of the system have to be weighed against the slightly greater costs of the aluminium container and in some cases the need to process in counter-balanced retorts to prevent the light gauge plate from deforming during thermal processing.

Since the mid 1970s tapered two-piece tinplate and aluminium have been available. Here the can body is drawn from a blank and transported while nested with other cans in tiers. The system offers savings in equipment, labour costs and space when compared with un-nested conventional three-piece tinplate cans; it also overcomes the need to complete fabrication in the cannery. as occurs with cans bodies that are despatched in the flat for later erection (reforming) and addition of the can maker's end, prior to normal filling and sealing.

Double Seam Formation and Inspection Procedures

The double seam is an hermetic seal formed by interlocking the can body and the can end during two rolling actions. The first action roll curls the edge of the can end up and under the flange of the can body and folds the metal into file thicknesses (seven at the Side-seam) while embedding the flange into the compound. During this operation the circumference about the edge of the can end is reduced causing the "extra" metal to wrinkle.

The second action roll flattens and tightens the seam so that an hermetic seal is formed. This action causes the wrinkles (formed in the first operation) to be ironed out while the compound is forced into any gaps between the metal surfaces. The various stages in the formation of a can double seam; in Figures 10 and 11 are a cross section of a double seam and the major attributes affecting seam quality.

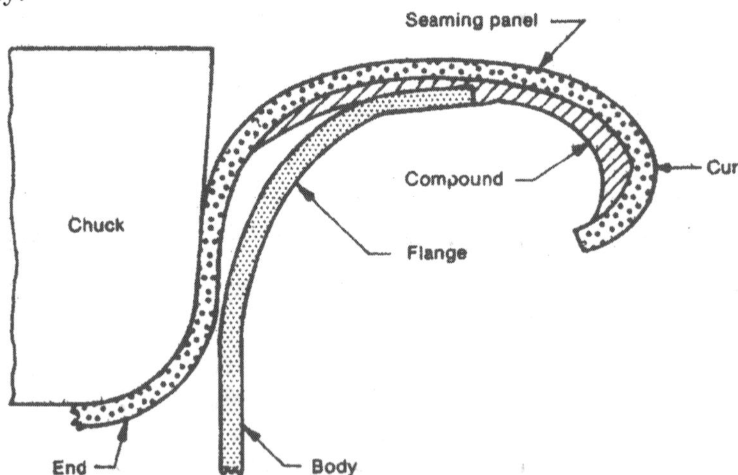

Figure: *Cross-section showing the positioning of the parts of the can body and loose end which will form the double seam (Courtesy of Standards Association of Australia.)*

As product safety depends upon maintenance of the hermetic seal, it is important that double seam formation be checked regularly during production, after all jams under the sealing machine, after adjustment to the machine, and after machine start-up following a long delay in production. Good manufacturing practice guidelines indicate that visual inspection of double seams should be at least every 30 min, while full tear down procedures should be followed for each sealing head at least every four hours. Can manufacturers and can seaming machine suppliers usually supply directions for seam formation and standards against which double seams are evaluated.

As a guide the major quality criteria for assessing double seam quality are summarized:

a. External inspection: much information as to the quality of a double seam can be obtained by a visual and tactile examination of the rolled seam. For skilled operators it is often not necessary to strip a double seam and measure the component in order to determine whether the sealing machine is rolling seam which comply with the requirements of good manufacturing practice.

   Conversely, an alert machine operator can pick up drift in performance before double seam criteria fall below acceptable limits. when conducting these assessments it is necessary to check for the following defects:

   - droop at the juncture spurs and skidders;
   - seam cut over, seam fracture and cut seam;
   - false seams (a point on the seam where the end and body hooks fail to engage);
   - damage to the double seam or can body.

   Also shown in Figure 13 are the locations where double Seam measurements of the stripped seam components of circular cans should be conducted, and in Figure 15 are shown the positions for component measurements on rectangular cans.

b. Tear down inspection: a complete analysis of the double seam form and dimensions; should be completed at least every four hours of continuous production for each seaming head.

   At times when there are difficulties with seam formation, these tests should be completed more frequently until satisfactory performance is demonstrated. The double seam attributes to be assessed include:

   - body hook butting (>70%),
   - overlap (>45%),

- tightness (>70%) ,
- juncture rating (>50%),
- countersink depth (>seam length at the same point),
- pressure ridge (continuous and visible).

In parentheses are shown the Australian recommended specifications for round cans of 74 mm diameter; however. as these values change for cans of different sizes and shapes. manufacturers ought consult their can suppliers to determine the satisfactory compliance criteria for use with their cans.

Sschematic diagrams for double seam sections of an end hook showing the juncture rating and the tightness rating. respectively. A cross section of a partially stripped seam in which the pressure ridge is visible.

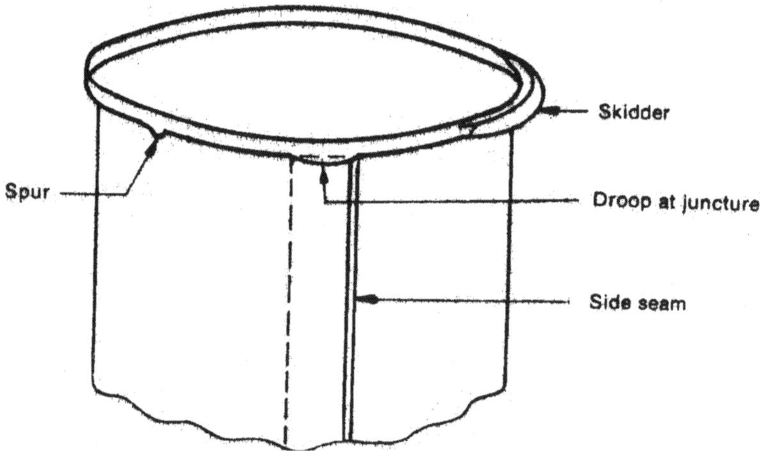

Figure: *Double seam Showing spur, a drop at the juncture and an incompletely rolled region known as a skidder (Courtesy of Standards Association of Australia.)*

Plastics and Laminates

With the development of plastic and plastic and aluminium foil, flexible, semi-rigid and rigid laminated packaging materials, has come a range of systems suitable for in-container sterilization of fishery products. Of these, the best known is the retortable pouch, which because of its flat profile and correspondingly high surface area to volume ratio (relative to that of cans), heats more rapidly than conventional cans. However, despite certain of their advantage,; (e.g., greater retention of heat labile nutrients and other quality, benefits arising due to rapid heat transfer to the thermal centres of retort pouch packs; the favourable costs of transportation, and the ease of opening and heating contents) they have not replaced, to the extent that was anticipated, conventional packaging materials for heat sterilized fishery products.

In America, problems were encountered with early versions of the flexible retort pouches which typically consisted of a three-ply laminate comprising an outer polyester layer for strength, scuff resistance and printability), a central foil layer (for excellent barrier properties) and an inner polyethylene or polypropylene layer (for heat sealability). These difficulties arose primarily because of FDA's concern regarding approval of the food contact surfaces used in flexible pouch manufactured and although they have been overcome, there still remain other disincentives arising from:

- the slow filling speeds (compare with metal cans);
- the difficulty of maintaining seal integrity when closing contaminated sealing surfaces;
- the difficulty of regulating the counter pressure required to assure a uniform profile during processing and cooling;
- the high cost of capital investment; and
- the need for protective outer wrappers.

Semi-rigid (all plastic), pouches (trays), are now available and with some of these systems the problems of slow filling and sealing speeds have been overcome by using integrated form-fill computer controlled equipment. Depending on the heat treatment selected fish processors may choose trays manufactured to withstand pasteurization conditions (i.e., at <100 °C) or sterilization conditions (i.e., at 110-122 °C). Irrespective of the form of the laminated container and the temperature at which it is processed, the function is the same -it must provide a strong hermetic seal, and because of this the seals should normally be at least 3 mm wide and continuous. For heat sealing the sealing surfaces should be plane-parallel to each other and the temperature of the jaw should be uniform across the entire sealing area. Since the integrity of the heat seal is critical to the safety of the product, it should be tested routinely. Typical testing protocols include,

- seal strength tests, normally used to determine the best combination of time temperature and seal pressure;
- burst-pressure tests;
- seal thickness tests;

- dye penetration tests; and
- visual appraisal of seal quality.

Whether considering retortable pouches which are flexible. semi-rigid or rigid. all offer the common attraction of providing a means to minimize the nutritional and sensory quality losses (which often are associated with traditional thermal processing in rigid containers), while simultaneously providing the opportunity to display visually appealing products. This is why developments with pouch packs are establishing a tradition of promoting a high quality image for fishery products.

Glass

With the exception of some fish pastes. glass is rarely used for fishery products which are preserved by heat alone; however. it is frequently chosen to package semi-preserved items such as salted fish. pickled herrings and caviars. The principles of processing in glass are substantially the same as for cans, but there are certain modifications which are necessary because of the sealing mechanisms used. and the thermal properties of glass, which make it vulnerable to rapid changes in temperature of more than 50 °C.

Sealing Mechanisms

Like cans, glass must be hermetically sealed to prevent product contamination after sealing and processing. Closures for glass container are made with either lacquered tin plate or aluminium into which has been placed a flowed-in plastisol lining compound (or a rubber ring with a pry-off cap) that acts as a sealant between the glass surface (called the "finish") and the cap. The closure is held in place by the vacuum in the container and/or the friction between the glass finish and the cap. The sealing surface of the glass may be across the top of the finish as with twist caps (Figure 19) or around the side of the finish as with pry-off caps (Figure 20) or around both the top and side seals as with push-on twist-off (PT) caps (Figure 21). It is important that the glass sealing surface be free of defects and protected from damage. as otherwise there is an unacceptable risk that the container will leak and draw in contaminants. It is because of the latter requirement that it is recognized as good manufacturing practice to ensure that the diameter across the finish of the jar is less than that of the diameter across the body of the container.

This prevents the closure from suffering undue damage through striking the closures on adjacent containers as they move along conveyors. Fortunately, with most containers that have lost their hermetic seals prior to processing, the caps will fall off during retorting and thus alert operators to pack failure.

In addition to obvious loss of vacuum. other faults to be aware of when using glass include the following:

- cooked caps: usually caused by mis-alignment of lug type closures while passing under the sealing machine so that the lug sits on top of the thread rather than underneath it. Cocked caps are readily visible as part of the top of the closure is raised;
- crushed lugs: occur when the sealing machine forces the lug of a twist cap down over the thread, rather than engaging it correctly. while winding the closure down onto the finish;
- stripped caps: result when the cap is over-tightened so that the lugs strip and splay-out over the thread of the finish;
- titled caps: occur when pry-off and PT caps do not sit down uniformly on the finish.

Inspection Procedures

The frequency of inspecting for adequacy of seals with glass containers should be sufficient to ensure consistent formation of hermetic seals. As a guide, this means that intervals between non-destructive testing should be no more than 30 min. while destructive testing should take place at least every four hours. In addition to this. visual inspection should follow every occasion that the capper jams. The results of all closure examinations should be recorded on the Appropriate form.

9

Present Status and Future Directives

In 1991 and 1992, over 700 000 metric tonnes (mt) of cultured shrimp were produced worldwide, with 80% of production coming from Asia. Production was expected to increase in succeeding years with the opening of new farms and adoption of more intensive culture methods. This expectation was not met, however. Global shrimp production in 1993 was only 639 000 mt, about 12% lower than the 1992 production. Crop failure in China in the summer of 1993 greatly affected the world's farmed shrimp production.

The increase in production seen in 1994 was mainly due to record crops in Thailand, India and Vietnam (Rosenberry, 1994). This production, however, still did not surpass the harvests obtained in 1991 and 1992.

Black tiger shrimp, *Penaeus monodon*, is the main species cultured in Asia except in China and Japan, where the main crops are *P. chinensis* and *P. japonicus*, respectively. Other penacids grown in farms are *P. semisulcatus, P. penicillatus, P. merguiensis* and *P. indicus*.

The phenomenal growth of the shrimp farming industry in less than two decades has created various problems. Disease epizootics of economic significance are a major constraint to the industry, not only because they affect the quantity of harvest, but because disease also affects the quality and regularity of production. At present, the main goal of the shrimp industry is to meet the growing demand in a sustainable manner without damaging the environment. The role that shrimp disease research must play to attain such a goal is now being seriously considered by both the private and government sector, and by national and international organisations. It is clear that collaborative effort involving all sectors is needed to attain sustainability. However, a sustainable industry can only be achieved if considerable investment in research is provided by many sectors.

Major Health Problems in Shrimp Culture Facilities in the Asia-pacific

Epizootics of both infectious and non-infectious etiology have continuously plagued the various sectors of the industry. Although it is generally recognised that intensive culture systems often encounter serious disease problems, more recent experiences have shown that low-density culture systems can also be severely affected.

A recent review on shrimp diseases has been made by Lightner (1993). In the context of Asian shrimp culture, excellent papers on diseases were presented at the *Workshop on Diseases of Cultured Penaeid Shrimp in Asia and the United States* (Fulks and Main, 1992). Diseases of both infectious and non-infectious etiology have been described, but their effects on shrimp and impacts on culture activities remain poorly understood. Diseases caused by viruses, bacteria, fungi and protistans are considered very significant to shrimp culture. The following list is largely based on Lightner (1993) and Lightner *et al.* (1994).

Diseases Due to Infectious Organisms

- *Viruses*: At least 15 viruses are known to infect cultured and wild marine penaeid shrimp. Reported types include parvoviruses, baculoviruses, reoviruses, togaviruses and rhabdoviruses. Three systemic baculoviruses have recently been described for penaeid shrimp: yellowhead virus (YBV) (Boonyaratpalin *et al.*, 1993), hemolymph baculovirus (Owens, 1993) and systemic ectodermal and mesodermal baculovirus (SEMBV) (Wongteerasupaya *et al.*, 1995). YBV and SEMBV infections have caused drastic mortalities resulting in severe economic losses in shrimp culture facilities in Thailand, Indonesia and India. Also described is a rod-shaped nuclear virus of *Penaeus japonicus* (RV-PJ) (Inouye *et al.*, 1994) which has been implicated in mass mortalities of cultured *P.japonicus* in Japan in 1993.
- *Bacteria*: Due to the economic losses following epizootics, bacteria are considered the most economically significant disease agents of shrimp. A review of the bacterial diseases

affecting different aquaculture commodities in Asia is presented by Lavilla-Pitogo (*in press*). Infections due to *Vibrio* spp. affect shrimp in hatchery and growout facilities, causing chronic or mass mortalities and shell deformities which affect marketability. Although bacteria are easily isolated from diseased shrimp, the mechanism for pathogenicity remains unclear. The course of action taken by most farmers to arrest bacterial infections is to use antibacterials. These products have been used indiscriminately, creating resistance problems. Issues, problems and constraints regarding the use of chemicals in aquaculture facilities were raised during a workshop at the *Symposium on Diseases in Asian Aquaculture* held in Bali, Indonesia in 1990 (Shariff *et al.*, 1992).

- *Fungal, protistan and other parasitic diseases*: Fungi like *Lagenidium, Haliphthoros* and *Sirolpidium* caused epizootics in shrimp hatcheries (Baticados *et al.*, 1990), resulting in heavy mortalities in larval stocks within two days. Infections due to the parasitic microsporidian *Agmasoma penaei* have been reported to cause high mortality in cultured shrimp (Flegel *et al.*, 1992). Protistans, such as gregarines and ciliates, are also associated with shrimp, but their general impact is less because of their relatively slow development.

Non-Infectious Diseases and Syndromes

Several nutritional diseases and syndromes have been reported in shrimp, such as chronic soft-shell syndrome (Baticados *et al.*, 1986) and red disease (Lightner and Redman, 1985).

Various nutritional, toxic and environmental diseases are discussed in Sindermann and Lightner (1988), Baticados *et al.* (1990), Flegel *et al.* (1992) and Limsuwan (1993).

Reports of mortalities due to toxins and pesticides are emerging. Flegel *et al.* (1992) describe the toxic effects of methyl-parathion and cypermethrin pesticides used in rice culture on postlarvae of *Penaeus monodon*. The emphasis placed on studying these pesticides is due to the fact that they can find their way into shrimp farms through run-off.

Environmental Problems

Problems resulting from the effects of shrimp culture on the natural environment have been raised (Primavera, 1993). Moreover, concern is also directed toward the effects of shrimp culture practices on the pond environment and towards the conflict among users of the same water source for various purposes. More detailed discussions are given by Phillips (this volume).

The Importance of Research

In the past, high production returns from shrimp farming were recorded in areas where new farms with good water source were opened. Opening of new areas is now limited, however, and future production will largely depend on already utilised resources. Successful production ventures will now largely depend on effective management utilising technologies developed through meticulous research.

As with fish pathology, the classic approach to shrimp pathology is descriptive in nature. Most published studies on shrimp disease deal with the isolation and morphological or biochemical characterisation of the pathogen. Recently, these activities have been carried out through molecular methods using gene probes. For effective disease control and surveillance programs, the following components should be considered:

Diagnostics

The use of new gene probe technologies that rely on demonstrating specific nucleic acid sequences offers an opportunity to detect viruses and other pathogens at much earlier stages of infection and to study how particular pathogens are transmitted to and within the shrimp host. Shrimp health status certification can also be performed with greater confidence in the outcome.

Epidemiology

The study of the distribution and determinants of health and disease in the natural setting is analogous to the pathogenesis of disease in individuals (Meek and Martin, 1991). It is clear that the classic approach to fish pathology should be augmented by epidemiology.

Epidemiological surveys, which are essential for managing the spread of shrimp disease agents, need urgent attention. The availability of standardized diagnostic methods with predictable results will make surveillance and diagnostic work easier. The epidemiological approach calls for cooperation between farmers and pathologists in data-gathering to assess situations pre-empting a disease outbreak. These data will be the basis for rational decisions for the prevention and/or control of disease.

Tissue Culture Establishment

Although this work is expensive and difficult, cell lines are important because they allow *in vitro* study of host-pathogen interactions.

Immunologic Studies

For effective diagnosis and control of infectious shrimp diseases, sufficient knowledge of immunology is needed. The latest information and recommended research strategies for marine molluscan and crustacean immunology are discussed by Bachere *et al.* (1995). They emphasize the importance of immunology as a tool for studying host-pathogen interactions at the cellular and molecular levels, and as a key to selection of pathogen-resistant animals.

Toxicology and Drug Efficacy Evaluation

The need for approved chemotherapeutants in shrimp culture is obvious. However, the approval process required is complicated, time-consuming, and expensive. Bell (1992) thoroughly discusses the issues related to drugs for shrimp diseases. Equal to the need for approval is effective enforcement of laws regarding the use of drugs in shrimp culture. Collaborative efforts with other disciplines will also lead to better shrimp health management. Husbandry techniques to produce fry and broodstock of good quality, as well as better nutrition and feeding strategies, will greatly support hatchery and farming activities. On the farm, environmentally friendly water and soil disinfection strategies are needed to get rid of pathogens in the water source from over-subscribed areas.

Effluent clean-up schemes need to be devised to reduce nutrient discharge and prevent eutrophication in receiving waters. Mialhe *et al.* (1995) also emphasize that beyond the strategy of prophylaxis based on diagnostics for preventing and managing epidemics, another strategy must be developed for selecting pathogen-resistant strains.

The Importance of Collaboration in Shrimp Health Research

As a field of research, shrimp pathology has limited manpower compared to disciplines like breeding, seed production and husbandry. Solutions to production constraints due to infectious and non-infectious disease agents are not easily met, thus creating a feeling of uncertainty among investors.

It is clear that to solve enormous problems with limited manpower capability, there is a need for cooperation and collaboration among institutions. The concept of an "international team" to work on problems related to infectious diseases in marine invertebrate aquaculture was put forward by Mialhe *et al.* (1995) to avoid duplication of work. There is also a need for increased interdisciplinary research. Pathologists can work with nutritionists in determining the nutritive values of new feed ingredients or the effects of anti-nutritional factors in these substances.

The role of scientific societies in carrying out plans for collaborative research is also discussed by M. Shariff (this volume). In Asia, the Fish Health Section of the Asian Fisheries Society serves as an effective mechanism for the exchange of ideas in research. However, private sector participation in this exchange is minimal. Research institutions, universities, government agencies and some private companies have contributed significantly to shrimp health research in Asia. Further strengthening of collaborative mechanisms to focus on shrimp diseases is still necessary to find fast track solutions to problems that beset the industry.

Effective collaborative projects have been initiated at the Aquatic Animal Health Research Institute (AAHRI) in Bangkok, Thailand with funding assistance from the Overseas Development Administration of the United Kingdom. The International Development Research Centre of Canada (IDRC) also funded research projects on shrimp disease at the Southeast Asian Fisheries Development Centre (SEAFDEC) and the Philippine Bureau of Fisheries and Aquatic Resources (BFAR). The role of universities in shrimp research is effectively carried out through the training of graduate students. This mechanism has boosted the manpower capability in shrimp pathology in Asia. Many universities and institutions in Japan, Taiwan, Malaysia, Australia and Thailand have contributed in training manpower in shrimp health research.

Recommendations for the Future

It is important for research to come up with information to secure the stability, sustainability and profitability of the shrimp industry. This could be done through:

- Developing rapid and sensitive methods to detect pathogens within the animal and in its environment;
- Increasing the availability and accessibility of disease diagnosis, prevention, and treatment services;
- Increasing the availability of quality fry for stocking (fry which are disease resistant, fast growing, tolerant to sub-optimum environmental conditions etc.);

- Developing a reliable method to assess fry quality;
- Increasing the availability of environmentally sound and cost-effective culture methods (determination of pond carrying capacity, fine tuning of feeding strategies, waste recycling methods, polyculture etc.);
- Undertaking studies on pond dynamics and its influence on shrimp health, and
- Developing or verifying technologies for the use of probiotics, bioaugmentation and bioremediation products for shrimp aquaculture.

Sufficient resources to develop and transfer technologies to support an industry faced with huge problems should be made available. Government funds, as well as contributions from international funding agencies, have initially funded research. Closer interaction with industry representatives should be made in order to come up with schemes to help finance expensive research.

Fundamentals of Hybridization in Fish Culture

Hybridization has become a common practice in fish culture, though it has not yet attained its rightful place in this branch of economy. The hybridization of fish for practical purpose has demonstrated its economic value. In the U.S.S.R. hybridization has solved the problems of carp culture in the northern regions. New productive forms have been successfully created by crossing species and varieties of Carassius, as well as by crossing carp and Carassius. Research on crossing of herbivorous fishes for their acclimatization in new areas has been initiated. Economic characteristics of a number of hybrids of ganoids have been determined. Experiments have shown that hybridization has extensive possibilities in sturgeon culture.

One may ask why hybrids should be created when excellent non-hybrids forms, sturgeon for example, are available. Hybrids are offered along with the pure species and not as substitutes. Preferences for hybrids are based on inherent characteristics of hybrid heterosis (hybrid vigour), which manifest as intensive growth rate, higher viability, adaptive flexibility, and sometimes early sexual maturation.

Charles Darwin attached great significance to hybridization: "If we do not take it for granted, at least we shall consider it highly probable - the existence of a great law in nature, the law of crossing animals and plants which are not closely related to each other, and this is extremely useful and even necessary".

Natural Hybridization

Study of causes and regularities of hybridization processes occurring in nature reveals possibilities and methods of utilisation of hybridization if fish culture. It should be noted that in no other group of the animal world but fish do remote (interspecific and intergeneric) crossings occur so frequently in nature, producing numerous hybrids. This is particularly true of fresh water fishes, as reproductive isolation is violated more frequently among them than among sea fishes because of less constant hydrological conditions in rivers as compared to the seas.

According to the literature available to us, 212 different fish hybrids have been recorded up to 1956, of which only 30 are of sea fishes. The number of hybrids in nature may be much higher, but not all of them are known to investigators or identified correctly.

In nature, hybrids occur mostly as single specimens, but sometimes large scale hybridization may occur as a result of violation of reproductive conditions of the species. Abundance of individuals of one species occurring along with less numbers of other species may result in an increase of interspecific crossing. That is why the probability of crossing is high when large numbers of a species are introduced into a new area, while it remains low in the aboriginal species.

Fish hybrids are of interest from two points of view:

i. As objects for commercial fish culture when heterosis of the first generation is to be used; sterile hybrids are suitable for this purpose; and
ii. As initial material for producing new hybridogenic forms when the ability for reproduction is a necessary condition. The problem of hybrid reproduction is the most important problem in hybridization.

Great variations in the reproductive ability of hybrids are observed: from complete fertility of intergeneric hybrids of some Cyprinodontidae to complete or incomplete fertility of interspecific and intergeneric hybrids of the sturgeon (Acipenseridae), the carp (Cyprinidae) sunfishes (Centrarchidae) and others.

The degree of fertility of hybrids does not always correspond to the degree of taxonomic relationship of the species crossed. In fact, intergeneric hybrids may be more fertile than intrageneric ones. Sometimes, even reciprocal hybrids produced by crossing the same species, but by substitution of a female of one species by a male of another, and vice versa, may be different in fertility. Individual hybrids may vary in fertility. In fact, some individuals of so-called sterile hybrids can produce offspring. Therefore, one of the methods of eliminating complete sterility in such hybrids is to carry out hybridization on a larger scale to provide a larger number of hybrid individuals, some of which may be fertile. Fertility may be expected to increase in the second hybrid generation. This is true of such hybrids as Cyprinus x Carassius which, as a rule, are sterile; but among them some exceptional individuals may be fertile. Investigations carried out by A.I.Kuzema has shown possibilities of successful selection work on Ukrainian hybrids of Cyprinus x Carassius.

Fertility of most hybrids is disordered to a certain extent, and they are sometimes sterile. In the process of evolution of various species, they became gradually isolated ecologically, geographically and in other respects. Due to isolation a sort of barrier appeared which prevented them from free crossing. However, such an isolation was not always reliable; the possibility of interspecific crossing still remained. This resulted in the development of a second barrier - disorder in hybrid fertility, preventing them from mixing and destroying specific differences. But where the first barrier acts without failure, fully eliminating the possibility of two species crossing, the second barrier might not appear. If such species are crossed artificially, the first barrier is removed, and a fertile hybrid is produced. Charles Darwin meant these very cases when he wrote "there exist species which are not easy to cross, but their hybrids, once produced, are highly fertile". Our experiments have shown that man, having succeeded in crossing such species, may find access to reproductive potentials that do not occur in nature.

Artificial Hybridization

It has been possible to produce fertile hybrids of beluga (Huso huso) and sterlet (Acipenser ruthenus), two species which do not cross in nature due to marked ecological differences. The sturgeons are remarkable for their ability to cross and for their hybrids that are fertile to various degrees. For many years we crossed the sturgeons in most specific combinations in order to produce interspecific and intergeneric hybrids, not only in the first generation, but also in the second and even the third generation (also back-cross and triple-cross). In our work we did not come across any instance in which it was impossible to produce viable offspring. Without detailed cytogenetic analyses of the causes of fertility and sterility of fish hybrids, I would like to point out that there are more chances of producing completely fertile hybrids if the species crossed have equal chromosome number, which must be homologous and able to conjugate in the course of gametogenesis. When there is unequal chromosome number and violation of the conjugation process, sterility occurs in various degrees.

Until recently the sturgeons and their hybrids were not studied cytogenetically. We have initiated research work on chromosome complexes of parent species and their hybrids, which has already given interesting results. The various species of the sturgeon family such as Huso huso, sterlet (Acipenser ruthenus), starred sturgeon (A. stellatus), ship sturgeon (A. nudiventris) and, probably, some others are characterised by a small number of chromosomes (about 60), whereas the chromosome number of A. guldenstadti is more than twice as much. Therefore, crossing of the first four species in any combination is likely to produce fertile hybrids; crossing each of the four with A. guldenstadti produces hybrids which are either of limited fertility or are completely sterile, the results of abnormal gametogenesis.

Hence, before starting work on hybridization, it is necessary to study chromosome complexes of the species to be crossed; this will permit prediction, with a greater degree of probability as to whether or not the hybrids will be fertile or sterile, and will avoid waste of labour and money.

Heredity of Hybrids

Many scientists, I.V. Mitchurin in particular, have shown that development of some characteristics of hybrids, to a great extent, depends on the environment in which the hybrids were bred. This environmental selection of characters is possible due to the great heterozygosity of the hybrid.

Study of morphological characteristics of certain hybrids of the sturgeons, the carps and others (as compared to their parent species) shows that hybrids have an intermediate heredity not only in the first generation, but also in the second, as well as in the generations produced by back-crossing and triple-crossing. In comparison to the first generation, the following generations, often but not always, are characterised by greater variability; however, no typical morphologic segregation followed by return to the initial species has been observed. Hence it was believed that interspecific hybrids possess the so-called permanent intermediate heredity, which does not obey Mendel's law of segregation. But recent experimental research has proved that the type of heredity observed in crossing different species does obey the law of segregation. But the law, revealed by the behaviour of genes, is impossible to observe visually in morphological characteristics.

The species crossed are different in many characteristics, and consequently in a great number of genes. In interspecific crossing high polymeric inheritance occur when numerous genes affect the development of some definite characteristics in a similar way. This is why in the second generation a much greater number of combinations is inevitable; and among the mass of intermediate specimens, only a few of them are homozygous, which, in all characteristics show a return to the parental form. However, often such individuals are not found in the experiment.

In intraspecific crossing, (inter-racial, for example) it does not matter which race is represented by the male or female, since the characteristics of the offspring are not affected. That is to say, reciprocal hybrids do not differ from each other. This is expressed by the formula: $A \times B = B \times A$. In interspecific, intergeneric and more remote crossings, reciprocal hybrids may be different, as for example the mule and the hinny which are the reciprocal hybrids between horse and the donkey.

Differences in reciprocal forms of some fish hybrids are revealed in the following:

(i) viability,

(ii) growth rate,

(iii) rates of development of other characteristics of heterosis and

(iv) morphological characteristics.

Marked difference in viability is observed in reciprocal intergenetic hybrids of crucian carp (Carassius carassius); the offspring of Carassius females crossed with males of other genera of Cyprinidae (for example Carassius s × A Tinca u) are viable, whereas the reciprocal offspring (of Carassius males) are not viable. Considerable differences in growth rates of fry were found in reciprocal hybrids of the sturgeons. The hybrid possessed some features characteristic of the maternal species, revealing matroclinic inheritance. Morphologic differences between reciprocal hybrids are even more pronounced: in countable characteristics (number of vertebrae, fin rays) these differences are again matrilinear.

Differences of reciprocal forms of hybrids are determined by cytoplasmic differences of the species crossed, resulting in different interaction of the nucleus and the plasma in reciprocal crossing. Genetic predetermination of cytoplasm may lead to incompatibility of the cytoplasm in certain species with chromosomes of some other species. As a consequence, some reciprocal hybrids may be viable, whereas some others may not be. The same is true of the factor determining the manifestation of various degrees of heterosis in reciprocal hybrids. Cytolplasmic heredity is a basis of matroclinic inheritance as well.

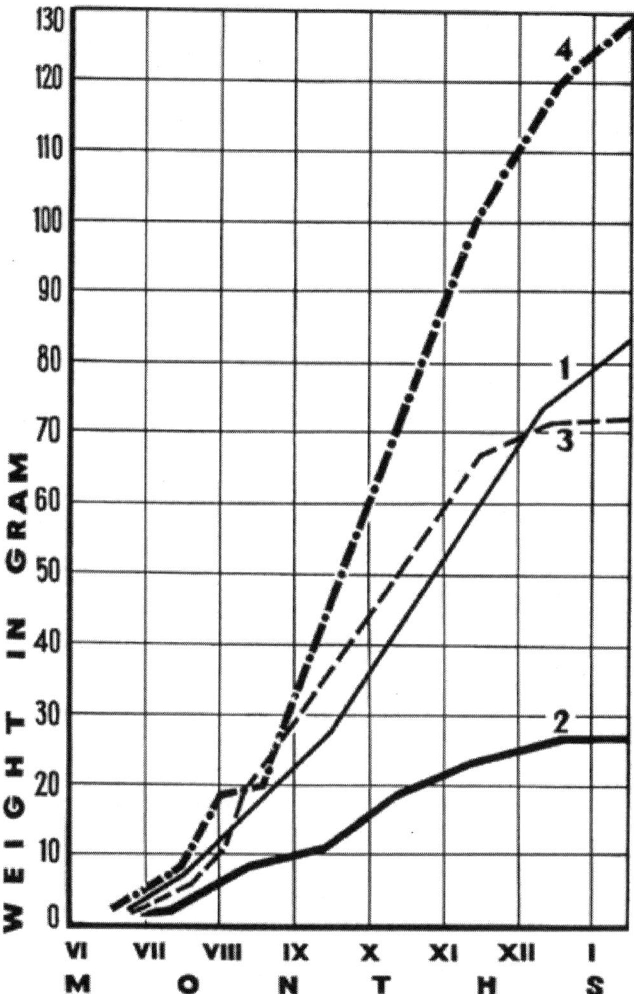

Figure: *Growth rate of reciprocal hybrids and parental sturgeon species in aquaria.*

1 Acipenser güldenstädti,

2 Acipenser ruthenus,

3 A.ruthenus s × A. güldenstädti u

4 A.güldenstadt. s × A.ruthenus u

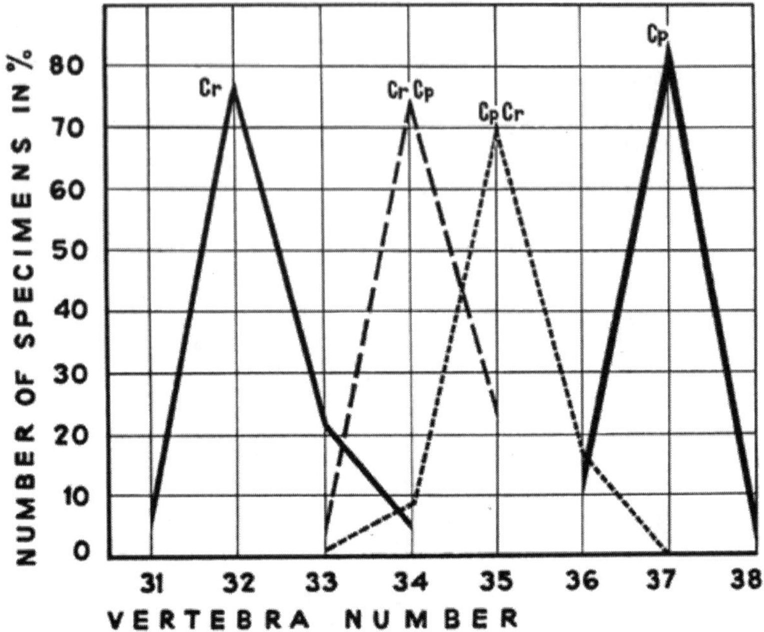

Figure: *Matroclinal differences in number of vertebra between reciprocal hybrids.*

Cr	-	Carassius carassius,
CrCp	-	C.carassius s × Cyprinus carpio u,
CpCr	-	C.carpio s × C.Carassius u
Cp	-	C.carpio

The nature of many phenomena in remote hybridization cannot be explained without consideration of gynogenesis as one fundamental problem of the theory. Gynogenesis is a form of sexual reproduction, when in the process of fertilization, the sperm penetrates the ovum and activates it for development without contribution of paternal chromosomes, so that heredity in the offspring is determined by the female pronucleus alone.

As a result, pure matroclinal pseudohybrids are produced, represented as a rule by females only. Some scientists consider gynogenesis to be a special case of parthenogenasis. This is not correct because in parthenogenesis, activation of the ovum is not initiated by sperm but by some other agents. Gynogenesis occurs in nature and can be demonstrated experimentally. Most interesting instances of natural gynogenesis are Carassius auratus gibelio and the Mexican viviparous fish Mollienesia formosa. They reproduce gynogenetically, spawning with males of some other species. Some experiments on remote hybridization have resulted in so-called hybrid gynogenesis. Our numerous experiments in crossing Carassius carassius females with Leuciscus cephalus males (different subfamilies) have shown that only a small part of the offspring was viable, and these were matrilinear, i.e. gynogenetic due to the exclusion of the male chromosome complex during development. Most of the embryos died at the very beginning of development owing to incompatibility of maternal and paternal hereditary substance.

Of great theoretical and practical interest, is radiation gynogenesis, produced when eggs are fertilized by sperm exposed to X-rays in dosages sufficient to stop its hereditary function, but not its ability to penetrate the egg and activate its development. In fact, in the fertilization of the sterlet (Acipenser ruthenus) eggs by the sperm of the sturgeon exposed to X-rays, or the beluga (Huso huso) eggs by the irradiated sperm of the sterlet, completely matrilinear offspring were produced (no paternal characteristics were observed). Since gynogenesis produces only females it may be possible to utilise it for control of sexual ratio in fishes, which is of great practical importance. Increase in the number of females of the sturgeon, for example, would be desirable both for commercial fishery and for artificial culture. Experiments in this field should be continued on a larger scale.

Heterosis

Heterosis in fish hybrids is a natural phenomenon observed in various families such as Acipenseridae, Salmonidae, Cyprinidae, Percidae, Centrarchidae and Poecilidae. As a rule heterosis is more prominent in intergeneric hybrids which are not closely related, than in interspecific or intraspecific hybrids.

Heterosis is revealed in various degrees. For example, growth rate of the hybrid may exceed that of the parent species (typical heterosis), or sometimes growth rate of the hybrid may exceed that of one of the parent species and be equal to or less than the other parent (incomplete heterosis).

In the second hybrid generation, heterosis usually declines, but in back-crosses of the first generation hybrid with the parent species (in which the characteristics that are of interest to us is more pronounced) the hybrid may exceed both the parent species; that is, the hybrid will show a prominent heterosis. For example, the hybrid of Huso huso x (Huso huso x Acipenser ruthenus) exceeds Huso huso in growth rate. Both in evolution and selection, hybrid forms are of more value if their fertility is not violated. Heterosis should be used on a greater scale in selectional culture of new, valuable fish and in acclimatization as well as in crossing of hybrids for commercial culture in ponds and other inland reservoirs.

Hybridization of Acipenseridae and its Practical Significance

Natural Hybridization

Sturgeons (Acipenseridae) are known to interbreed under natural conditions, giving rise to viable and sometimes fertile interspecific and intergeneric hybrids. Hybrids have been described from crosses of various combinations of almost all species of the family. L.S. Berg in his well-known book "Freshwater fishes of the USSR and the neighbouring countries" (1948) enumerates the hybrid forms of this family from the following crossings:

1. Huso dauricus × Acipenser schrencki
   (kaluga × Amur sturgeon)
2. H. huso × A. nudiventris
   (beluga × spiny sturgeon)
3. H. huso × A. güldenstädti
   (beluga × sturgeon)
4. H. huso × A. stellatus
   (beluga × stellate sturgeon)
5. A. nudiventris × A. stellatus
   (spiny sturgeon × stellate sturgeon)
6. A. ruthenus × A. güldenstädti
   (sterlet × sturgeon)
7. A. ruthenus × A. stellatus
   (sterlet × stellate sturgeon)
8. A. güldenstädti × A. stellatus
   (sturgeon × stellate sturgeon)
9. A. baeri × A. ruthenus
   (Siberian sturgeon × sterlet)

This list does not include the whole variety of acipenserid hybrids that can occur in nature, as for instance the hybrid between beluga and sterlet (H. huso × A. ruthenus). The capacity of Acipenseridae to interbreed resulted in higher variability and formation of different varieties.

Artificial Hybridization

The first artificial hybridization of Acipenseridae was accomplished by F.V. Ovsyannikov, who as early as 1869 fertilized the eggs of sterlet (A. ruthenus) by the sperm of sturgeon (A. güldenstädti) and stellate sturgeon (A. stellatus), and drew attention to the importance of research in this direction. Later, however, experiments on hybridization of Acipenseridae were discontinued for a long time. Investigations were resumed in 1949 in the Volga River, in line with the research programme of VNIRO (All-Union Research Institute of Marine Fisheries and Oceanography).

The task set was to combine in the hybrid organisms the properties of such large anadromus Acipenseridae as the beluga (H. huso) and the sturgeon (A. guldenstädti) with the properties of the freshwater sterlet. It was assumed that products of the cross between sterlet and sturgeon or beluga, with increased adaptive properties and heterosis, would be distinguished for their economic value as compared to the parental species. Since hybridization involved the sterlet, which attain sexual maturity earlier than other Acipenseridae, it was assumed that the hybrids produced would differ favourably in this respect from late-maturing anadromous Acipenseridae. Experimental research confirmed the soundness of these suppositions. It is expedient to make use of the heterosis of the hybrid forms of Acipenseridae when rearing them in freshwater ponds and reservoirs.

Pond Breeding of Hybrids

Though Acipenseridae are suited for commercial breeding in ponds they have seldom been used for this purpose in the past. Sterlet were successfully bred in ponds for many years, but this practice has now been nearly abandoned because of the slow rate of growth of this fish. Of interest in this respect are some hybrid forms which adapt themselves more easily to pond conditions that are unusual and hardly suitable for Acipenseridae. Besides, the favourable characteristics of hybrids based on heterosis, viz., rapid growth, greater viability, and early sexual maturity may be advantageously used. Among the acipenserid hybrids studied, the most promising in respect to fisheries is the hybrid between beluga and sterlet, which favourably combines the valuable properties of both parental species: the rapid growth rate of beluga and the early maturity of sterlet.

Growth

Even in ponds of low fish productivity, where the potential speed of growth of the hybrid could not manifest itself to the full, it reached the weight of 500 g by the end of the second summer, while sterlet under the same conditions grew much slower. In one of the ponds of the Teplovsk fish hatchery (Saratov region) which has greater food reserves, the hybrid fingerlings reached the mean weight of 74 g in one summer, the yield of fish being 580 kg/ha. In the other ponds where the hybrids were cultured along with carp, the fish production did not exceed 300 kg/ha, the growth of the hybrid being slower as they could not compete successfully with carp for benthic food.

The potential rate of growth of hybrids is quite high when sufficient food is available, one-summer fish growing to an average weight of 0.5 kg. This potential for rapid growth can be used advantageously by rearing them in ponds to marketable size with the help of intensive artificial feeding. Within two growing seasons it is possible to produce marketable hybrids weighing nearly 1 kg. Since 1966 the efficiency of breeding marketable stocks of hybrids in some fish farm ponds of the Soviet Union is being reviewed. For this purpose VNIRO has worked out "Directions on pond breeding of marketable hybrids of beluga and sterlet by intensive feeding" (1966).

The directions give special attention to the bio-techniques of feeding young hybrids and to the adverse effects of growing carp and hybrids together; it is also pointed out that fish ponds should not be overgrown with filamentous algae.

Food

The favourite food of very young hybrids is larvae of Chironomidae, called bloodworms. Later, those that have inherited predatory instincts from beluga start feeding on larger animals such as tadpoles, small frogs and any small inactive fish. Fingerlings of hybrids can attain average weight of 90–100 g in ponds with abundant food reserve, whereas in the sea, where the standing crop of benthos is many times higher than in ponds, e.g., in the Taganrog Bay of the Azov Sea, they average 500 g. Besides using benthic invertebrates, hybrids in the sea feed on numerous slow-swimming bottom fish, mainly Gobiidae, which are not to be found in ponds.

To supply Acipenseridae with a sufficient quantity of food in ponds, supplementary artificial feeds should be introduced on feed tray. Since all Acipenseridae in nature feed only on animal food they should be fed on food of animal origin too: scrap fish or fish of little value, wastes of fish and meat processing plants, various invertebrates such as Gammaridae, etc. It is desirable that all these products be given fresh or frozen, and if this is impossible, these feeds can be preserved with sodium pyrosulphate. The conversion ratio (food per unit of weight gain of fish) may be assumed as equal to 7.

Stocking Rate

The stocking rate of the young may vary, approximately 15 000 to 20 000 individuals per ha, depending on the natural fish productivity of the pond and on the possibilities of regularly introducing supplementary food, starting, at least, from the second half of the growing period. The stocking rate of yearlings is 2 000 individuals per ha and two-year-old hybrids are fed intensively during the whole growing period. If the rate of stocking is higher feeding should assume still more importance, since the capacity for natural fish productivity of ponds decreases. The rearing practice should then acquire the character of tank fish culture.

Yield

To raise fish production in ponds, it is advisable to rear hybrids together with herbivorous fishes of the family Cyprinidae, provided they are of the same age group. The stocking rates of hybrids, mentioned above, may be retained, but it is desirable to lower the usual stocking rate of herbivorous fishes by approximately 30 percent.

In the autumn of the second growing season, two-year-olds weighing not less than 800 g are delivered to the market. A number of hybrids that have not reached marketable size are left to grow in the pond for a third summer. Due to increased viability, hybrids winter well in the usual carp ponds; the mortality rate, even among one summer fish, being quite insignificant. The tentative biological standards, worked out by VNIRO and tabulated below, may serve as a general guide for producing a marketable crop of hybrids of beluga and sterlet.

Table: *The table presents minimum yield indices which may be obtained even in non-specialised carp-type fish farms.*

Rearing time	*Stocking rate per ha*		*Yield*			*Fish productivity kg/ha*
	Number in thousands	Weight (g)	Percent	Number in thousands	Average weight (kg)	
First summer	20	3	50	10	0.08	800
Second summer	2	80	80	1.6	0.8	1 150

By way of illustration the results of rearing hybrids in the Aksay fish farm in the Rostov region are of interest. In 1965, pond no. 3, with area of 0.1 ha, was stocked with 3 700 young hybrids weighing 4 g each. The natural food resources were extremely poor; the standing crop of benthos averaged 0.13 g/m^2.

The young were fed mainly on minced fish. The yield (survival) of young of the year, with average weight of 62 g, was 58.4 percent of the number of fingerlings stocked. Fish productivity equalled 1 330 kg/ha. Two-year-old hybrids in the farm averaged 0.9 kg the maximum individual weight being 1.8 kg.

In 1967 an experiment was undertaken in culturing mixed age groups of one-summer fish and three-year-olds. Pond No. 4 of 0.1 ha area was stocked with 1 000 fry of 4 g mean weight and 213 two-year-olds with mean weight of 700 g, the total stocking rate being 12 130 individuals/ha. As a result, fingerlings of average weight of 85.2 g (yield - 72 percent), and three-year-olds of average weight of 1.94 kg (yield - 80 percent) were produced. Fish productivity of the pond was 570 kg/ha for fingerlings and 2 200 kg/ha for three-year-olds, the total productivity being 2 770 kg/ha.

Reproduction

Of specially great significance is the unimpaired ability of the hybrid of beluga and sterlet (different genera) to reproduce. This is a very rare phenomenon in intergeneric hybrids. The onset of sexual maturity in the hybrid is early; in males - at the age of 3–4 years; in females - from the age of 6–7 years. The size of eggs of the hybrid is nearly the same as that of the stellate, but smaller than Huso huso.

The fecundity of the hybrid and the high flavour qualities both of the flesh and of the caviar increase its value in commercial culture. The possibility of producing second and subsequent hybrid generations opens prospects of selective breeding of new varieties of Acipenseridae.

Thanks to heterosis and the heightened adaptive plasticity, the sexual products in the hybrid broodstock mature even when they are bred and reared in ponds throughout their whole life cycle, never entering a river or a sea.

The fish cultural problem of producing progeny of Acipenseridae has thus been solved, even raising them entirely under pond conditions. Here the positive role evidently has been played by the method of distant hybridization. However, numerous attempts to achieve this with thoroughbred Acipenseridae have not been successful so far.

Viability

The high viability and adaptability of the hybrid is illustrated by a batch of fish that for a long time (about 6 years) starved in the ponds to such a degree that they not only did not gain weight during the growing season, but many of them lost weight, could in the subsequent years when fed well produce both sperm and ripe eggs.

The unusual viability of the hybrid made it possible to use the method of stripping eggs, by incision, from live females that had received hypophyseal injection, after which the incised abdominal wall was sutured. The operation was well endured by the hybrid females. By this technique, I.A. Burtsev (in 1967) obtained eggs for fertilization and production of the second hybrid progeny from 5 females averaging 8.3 kg. The eggs from one female weighed 1 kg. The females which were operated on were returned to the pond of the Aksay fish hatchery. Examination of the females in October, when transplanting them to wintering pond, showed that they were in good condition; the sutures were completely grown over and hardly noticeable as pale stripes. The females not only survived but gained considerable weight during the growing season, from 1.7 to 3.6 kg. Thus the operation did not have any adverse effect on the subsequent condition and growth of the females. The technique of successfully obtaining eggs from living females, applied for the first time in sturgeon culture, might enable the repeated use of individual brood fish. This makes it possible to select females by the quality of their offspring, which must play a significant role in selection work.

Hybrid F_2

The second brood of hybrids of beluga and sterlet (F_2) was first obtained in the Rogozhinsk sturgeon hatchery, located in the Don River delta. The first was produced in 1966, and a considerably greater number (33 000 fry, reared to average weight of 3.5 g) was produced in 1967. The fish production indices (percentage of fertilized eggs, survival of young in reservoirs and ponds) proved in the majority of cases to be not lower, but at times even higher, than in the hybrid F_1.

Nearly all the young F_2 offspring were transported to the Don fish rearing plant in the Donetsk Region for pond culture of the young-of-the-year and for carrying out selection experiments. Some of the young produced from the eggs of each of the five females were introduced separately into five small ponds, the one-summer fish averaging 79.4 g, the maximum individual weight being 250 g. In rearing one-summer fish of hybrid F_2 and silver carp (Hypophthalmichthys molitrix) together, the total fish productivity of one of the ponds equalled 1 300 kg/ha. Morphologically, both hybrid F_1 and F_2 take an intermediate position between the parental species, but in fact there is a much greater variety in F_2 individuals. As compared to F_1, it especially varies in body colouration. The individual variability in F_2, of some countable traits and measurements, is also higher.

It is of interest to note that cytogenetically both broods resemble each other and the parental species: the modal number of chromosomes in all of them is 60, varying somewhat greater in F_2 than in F_1 and the parental species. This follows from the comparison of the variation sets. Consequently, the hybrid is characterised by a heightened variability both morphologically and cytogenetically (Nikoljukin, 1966).

The possibility of producing new breeds of the beluga x sterlet hybrid should find application in the production of new breeds with stable heredity, especially the pond breed of Acipenseridae. On the other hand, the possibility of producing progeny from hybrid broodstock reared under pond conditions opens prospects of organising specialised sturgeon farm ponds covering all the processes of fish culture: from obtaining fertilized eggs to producing a marketable crop without obtaining fry hybrid from sturgeon hatcheries.

Stocking of Hybrids in Reservoirs

Due to the wide range of adaptability of the beluga x sterlet hybrid, it can dwell in purely fresh waters ponds, or in brackish water bodies such as the Proletarsk reservoir, or in salt waters like the Azov Sea (Nikoljukin, 1964). In 1962 a hybrid was introduced into the Proletarsk reservoir, situated on the brackish Zapadny Manych River (tributary to the Don River). The zone of flooding of the reservoir's eastern part wholly covered the salt lake Gudilo, occupying over one-third of the whole reservoir. In late June the reservoir was stocked with 330 000 young hybrids with average weight of 7 g. In late August one and a half to two months after stocking, seine catches revealed that the young had survived and grown better than in ponds, the majority of hybrids having attained over 100 g.

In 1966, in the fifth year of life, many individuals attained the weight of 8 kg. This same year, mature males with ripe sperm were found. It may be expected that females will also start maturing in 1968. Though the purpose of the introduction of the hybrid into the reservoir was mainly to breed the fish as a marketable product, mature spawners may also be used for artificial propagation at the sturgeon rearing station in order to obtain new hybrid offspring.

Hydro-electric power projects on large rivers such as the Don, have strongly interferred with the reproduction of Acipenseridae, barring their way to the most important spawning areas. The efficiency of natural reproduction of Acipenseridae, the area of which is confined to the lower reaches of the Don, has gone down to such a degree that the progeny of beluga and sturgeon, in most cases, cannot be considered to be of any significance. Reproduction of Acipenseridae has become possible mainly due to their artificial breeging in hatcheries.

Introduction of Hybrids into Sea

Of great interest are the experiments being carried out at VNIRO for producing a new form of Acipenseridae by hybridization and acclimatization of this new form in the Azov Sea. The aim is to produce a hybrid with the rapid rate of growth of beluga and attaining maturity much earlier (Nikoljukin, 1964). The introduction of the hybrid of beluga and sterlet into the Azov Sea is worthwhile because the heterosis of the first generation may be used for increasing the fish productivity of the sea with the valuable Acipenseridae. An even more important task is to create a new, early-maturing form able to reproduce more efficiently in the Don below the Tsymlyansk hydroelectric power plant.

The basis for obtaining this new form (breed) is the hybrid between beluga and sterlet, the young of which have been released in the Don delta since 1963. From here the hybrids have spread into the Taganrog Bay and the Azov Sea proper, where they grow rapidly, predating on other fish. The survival rate of the hybrid is high. Thus, in the autumn of 1964 young hybrids prevailed in the commercial catches in the lower reaches of the Don and in the Taganrog Bay, constituting 50 percent of the total catch of young Acipenseridae. The young of the year feed mainly on Musidacea and fish, whereas two-year-old hybrids feed exclusively on fish (Gobiidae and Clupeonella sp).

Intensive feeding accelerates the sexual maturity of the hybrids. In the autumn of 1965, a 3 kg male aged 2+, with well developed tastes at maturity stage IV, was caught in the Don. This testifies that males become mature by about 3 years of age. In 1967, in the fifth year of life, a more intensive spawning migration of males from the sea into the river was observed. By this time the hybrids had attained the weight of 13 kg. Partial onset of maturity may be expected in 1968.

Lines of Future Work

When producing the new hybrid form, the following sequence of crossing is to be observed:

Beluga × sterlet (F_1)

Beluga × (beluga × sterlet) (F_B)

(Beluga × sterlet) × /beluga × (beluga × sterlet)/

The adult male spawners of the first hybrid brood are used for backcrossing with beluga females, whereas females of F_1 are crossed with the males of the backcross (F_B).

Of the three crossings, the first two (F_1 and F_B) have been repeatedly made by us, yielding positive results. The last (third) cross of the programme scheduled, will be made approximately in 1970–1971. In the progeny from crossing F_1 and F_B, the hereditary characters of beluga will prevail (5/8) over those of sterlet (3/8); and consequently they will evidently surpass F_1 in the rate of growth.

By the crosses mentioned above, the heterosygosity and heterosis of the hybrid forms will be maintained at a certain level. The hybrid form from the last (interhybrid) crossing should be propagated.

When producing a new form, selection experiments are aimed, above all, at rapid growth and early sexual maturity.

Levels and Methods of Genetic Resource Preservation in Fish

From the point of view of the preservation of the genetic resources of fish, four main levels of concern and strategy can be identified: (1) oceanic systems, (2) continental waters (both fresh and marine), (3) aquaculture and (4) stock enhancement programmes. The last two levels of concern are directly controlled by man, whereas the first two can be regarded as uncontrolled systems, although human activities can have major impacts on them.

Human Impacts on Aquatic Habitats

There are a number of characteristics which clearly differentiate oceanic systems from continental waters as far as genetic resource preservation is concerned. Firstly, the much greater size of oceanic systems, in terms of both physical dimensions and biomasses involved, make them much more difficult to manage. It must be remarked, however, that a few of the larger river basins (e.g., in South America and Africa) and their exploited fish biomass approach oceanic dimensions.

The effects of meteorological and especially hydrological cycles are usually much more crucial in continental waters where they have a strong influence on fish populations - especially as far as reproductive migrations or strategies are concerned.

The combination of these characteristics means that both natural and man-induced stresses have a relatively greater impact in continental systems. Included in this group are intertidal habitats, coral reef systems, brackish waters and fresh waters.

Table: *The scope of genetic resource preservation in fish*

Level	***Potential Genetic Problems***	***Methods***
Oceanic fisheries	Sub-population extinction	Tagging and monitoring
Continental habitats	Species extinction Genetic erosion	Establish reserves Scientific management of reserves Restocking
Aquaculture	Genetic erosion Inbreeding depression Loss of fitness	Controlled breeding Hybridization Cryopreservation
Stock enhancement	Genetic erosion Inbreeding depression Loss of fitness Introgression with wild stock	Controlled breeding Hybridization Cryopreservation Genetic sterilization of introduced stock

Biologically, these Conditions Give Rise to Two Further Distinctions Which are Relevant to Genetic Resource Conservation:

(a) continental areas have very much higher numbers of endemic and locally distributed species and sub-populations than oceanic systems, and

(b) the lower impact of stresses on pelagic oceanic species, and their high fecundity, make their extinction unlikely. At low abundance levels their exploitation will be uneconomical, whilst in enclosed continental systems many species could, and indeed have, disappeared.

In oceanic systems, the main changes are produced by the direct but selective exploitation through fisheries. Such exploitation can have two different effects on a considered species or population. One is a direct effect on the population structure of the target species by selective removal of a particular sector. The other is an indirect effect on non-target species, where the fishery may affect the food chain by, for example, upsetting established interactive relationships. In continental waters the same effects can occur but often to a much greater degree and whole populations can become extinct, solely through over-fishing.

Here, however, other important stresses occur, notably pollution, disruption of life history cycles and disruption of ecosystem structure. Examples include the introduction of foreign predators, competitors and pathogens. Among the different kinds of pollution, air-borne pollution leading to acid precipitation probably has had the most devastating (and sometimes selective) effect on fish populations particularly in southern Scandinavia and the north-eastern and north-central U.S.A. The direct pollution of waters through the discharge of various chemicals (some of them mutagenic) as wastes or the run-off of agricultural chemicals is also of major importance. Land-use may also have highly undesirable effects such as the production of very acid soils (and run-off) and the silting of streams. Another serious human impact is the disruption of biological (in most cases reproductive) cycles in fish populations as a consequence of building dams or other obstructions on rivers where migratory species occur. Dams may not only prevent the migration to upstream spawning grounds, but they may also change rivers into semi-lacustrine habitats quite unsuitable for stream species. Ecological barriers are also common and may be caused by zones of pollution in the lower reaches of rivers preventing the migration of various species.

The introduction of exotic species may also be included as a factor of importance to endemic fish populations. They may lead to the introduction of disease, higher levels of predation, or so affect the ecosystem (e.g., by competition for food) as to cause the extinction of local species. It is very rare to find introductions which have filled a totally empty niche, especially in the tropics, so introductions are highly likely to result in changes in the endemic populations. Compared to oceanic and continental systems, aquacultural, hatchery and stock enhancement programmes are subject to an even greater degree of human control by definition. Within such systems, direct environmental and genetic management can and should be imposed and should only be limited by economic criteria and technological sophistication.

Ways and Means of Genetic Resource Preservation

A strategy for genetic resource preservation is largely dictated by the type of aquatic system.

Oceanic Systems

For any oceanic fish species, one of the first requirements for the formulation of a preservation policy must be a distribution map and the identification of any distinct sub-populations which there may be.

Distribution data will probably be available from relevant fisheries, but the detection of sub-populations may have to be carried out by the use of one or more of the 'labelling' techniques available (e.g., tagging, the identification of meristic characters, electrophoresis, etc.).

The importance of detecting sub-populations is that, although the danger of the extinction of oceanic species through overfishing is probably very small (the assumption being that the fishery will disappear before the species), it is perfectly possible that a unique sub-population could disappear. (Incidentally, the practice of a rushed, published description of a local population as a new sub-species, merely for the purpose of forestalling development, is a legalistic device that will probably prove counter-productive, since it is likely to be abused.) Significant changes within an exploited population can only be detected through a continuing programme of monitoring designed to study the overall population structure, its composition and its change. The decline in any part of the population can probably only be controlled by regulation of the fishery, either through reduction of exploitation in space or time, or by altering the nature of the gear used.

The effect of fishing on the genetic diversity of oceanic populations is discussed below. If there is evidence of undesirable genetic changes in any species because of selection imposed by the fishery, it would be ideal to try to reverse the trend by appropriate management tools.

Continental Waters

One of the enormous problems related to smaller enclosed systems, especially in the tropics, is the identification of the species involved. In parts of Africa, Asia, and especially in South America, large numbers of endemic and very locally distributed fish species occur and many are undescribed. Thus an essential pre-requisite to any broad programme of genetic resource preservation is a proper taxonomic study of the fish species occurring in each area and a full check-list of these species, indicating the status of each, and, if possible, its significance in ecological, economic, scientific and social terms. Such lists would provide the basis of any international list of endangered species to be included in the Red Data Book published by the IUCN. Because of the high probability of the loss of genetic diversity in any species taken into culture, the best method of maintaining this diversity is to conserve self-maintaining populations in natural habitats. The normal procedure here would be to establish nature reserves (lakes, river basins, estuaries, coastal lagoons, coral reefs, etc.) which already contain one or more populations of the fish species or communities concerned. These reserves could well be multi-purpose areas, conserving several other types of habitat and communities in addition to the fish resource, e.g., National Park of Sabana Grande, Venezuela.

In some situations an individual species or its habitat may be so threatened that the only option is to collect stock and transfer this to an aquaculture system, or preferably, an alternative suitable location to create a new population. Ideally this new location should be within a nature reserve or in an area not liable to man-made pressures. The refuge so created can later be used to re-stock the original habitat if conditions there improve. This strategy is being used at present in both Canada and Scotland (Maitland, 1979) to preserve local populations of whitefish (Coregonus). Aquaculture techniques may be used on a short-term basis to enhance local stocks which are subject to temporary recruitment problems.

By definition, it is essential that man-induced stresses be strictly regulated in any established nature reserve. This includes control not only of fishing effort and gear, water quality degradation, barriers to migration, etc., but also of introduction of exotic fish or any other species likely to have a harmful impact on the ecosystem. The wild trout and charr watch concept, as recently proposed by Regier and Powers (1979), may prove to be one useful way of developing international monitoring schemes for important species or groups of fish. Once aquatic reserves are established, scientific management to prevent degradation and loss of diversity is absolutely essential. In terrestrial ecosystems, management of nature reserves is a new and often contentious discipline, but the high rate of extinctions for vertebrate species in terrestrial reserves requires the immediate attention of managers and consultants (Frankel and Soule, 1981). Fishery and aquatic reserve managers would be well advised to consider this problem at the outset.

General Principles of Genetic Resource Preservation

The Importance of Genetic Variation

Genetic variation is the raw material in species populations which enables them to adapt to changes in their environment. New genetic variation arises in a population from either spontaneous mutation of a gene or by immigration from a population of genetically different individuals. Alternate forms of a particular gene (or locus) are called "alleles". The number and relative abundance of alleles in a population is a measure of genetic variation, sometimes termed "heterozygosity".

Genetic variation is a measure of a population's ability to adapt to environmental change or stress, and thereby to survive.

Population geneticists have spent several decades establishing the importance of genetic variation in natural populations. It is known, for example, that the response to natural selection by experimental populations is accelerated by mutation-inducing radiation and/or the introduction of genes from different strains. In terms of the preservation of genetic resources, we can expect that benefits would be derived from maintaining the maximum level of genetic variation in a strain as well as by maintaining multiple strains that could serve as additional sources of genetic information by hybridization. It follows that the loss of genetic variation for whatever reason (e.g., prolonged selection, inbreeding, isolation) will result in a decrease of the potential adaptability of a population.

It appears that the benefits of multigene heterozygosity are universal in outbreeding organisms. In several organisms, including some fish species, individuals possessing the most genetic variation have been shown to have better survival rates or higher relative growth rates. Relatively heterozygous individuals appear to be more resistant to environmental perturbations during development. Clearly, genetically variable populations have many advantageous characteristics that are absent from genetically impoverished ones.

Over the past several years a rather large body of evidence has accumulated on the biochemical differences between alleles of genes coding for metabolically important enzymes. These biochemical differences emphasise the relationship between genetic and functional diversity. The functional properties of different alleles often reflect a biochemical and genetic adaptation to life in a heterogeneous environment. The extent, however, to which the states of physiological and behavioural traits at the whole animal level can be correlated with genetic and biochemical data is still uncertain. Most workers in the field agree that a proportion of protein polymorphisms have no direct or measurable effect on viability or some other aspect of fitness. Nevertheless, there is significant evidence from studies of individual genes, organisms and populations to substantiate the importance of genetic variation to population adaptability.

Effects of Inbreeding

The selection of small numbers of parents (Section 4.3) can reduce genetic variability. Equally serious is the fact that brood stock may be continually selected from closely related, perhaps full sib individuals.

This leads to generation after generation of inbreeding of closely related individuals which very often results in homozygosity for unfavourable genes. The overall result is inbreeding depression. Inbreeding depression is the loss of fitness (e.g. vigour, viability, fecundity) in connection with the loss of genetic variation due to homozygosity. The evidence that inbreeding is harmful is copious and virtually universal (Allendorf and Utter, 1979; Kincaid, 1976a, b; Kirpichnikov, 1972, and Kosswig, 1973).

A general view of the effects of inbreeding and its relationship to the conservation of genetic resources can be found in Soule (1980) and is treated more extensively by Frankel and Soule (1981). Quoting from the former: "A survey of inbreeding experiments leads to the generalisation that increasing the inbreeding coefficient by 10 percent induces a 5–10 percent decline in a particular reproductive trait." Note that an F value (inbreeding coefficient) equal to 10 percent approximates the amount of inbreeding that would theoretically occur in a population of five adults breeding at random for a single generation, or in a population of 25 adults breeding at random for five generations.

A 5–10 percent decline in fecundity might not appear to be very serious (especially when dealing with such fecund animals as fish) but if the effects of inbreeding depression on other traits (such as viability) are also considered this amount of inbreeding can lower reproductive potential as a whole by 25 percent (e.g., in fowl and swine). Gjederm (1974) showed that a 10 percent increase in the inbreeding coefficient in rainbow trout can result in a 10 percent decline in hatchability, and a 24 percent decline in viability of fingerlings. It should be noted that such independent effects are multiplicative in their impact on total, absolute survival and reproduction. Thus the unavoidable conclusion is that relatively small amounts of inbreeding can do tremendous damage to the reproductive potential and productivity of a fish stock.

Expected inbreeding depression is related to the current inbred state of a population. For example, a very small and isolated lake population of a certain fish species may be highly inbred because of its inherent demographic structure. The inbreeding of such a fish would not be expected to display an inbreeding depression as large as from a previously outcrossed group. In some breeding systems inbreeding is the goal and its associated "depression" is an undesirable but expected result for which compensatory breeding programmes can be utilised. Using quantitative methods of Nace et al. (1970) and assuming an average recombination frequency of about 0.1, Nagy et al. (1979) estimate that one gynogenetic generation is equal to 10–12 generations of full-sib mating. In many other breeding and brood stock selection programmes, practical logistic and economic management considerations can and have resulted in the inadvertent selection of brood stock in a way which causes close inbreeding. Unfortunately, once it has occurred, inbreeding depression is not reversible except by hybridization.

Monitoring and Measuring Genetic Variation

Technologies exist for a direct assessment of the genetic properties of a population. In some species the genetic basis of variation in some visible characters (e.g., colour patterns) can be established by breeding experiments, and the characters, or phenotypes, can be used to directly assess gene frequencies in populations. Traits such as colour patterns are typically controlled by a small number of genes (one to three). As such, those genes may not be representative of either overall genetic variation or population structure, but probably reflect fine-scale ecological or social structuring such as family or age-class recognition. Although these characters can be conveniently assessed in a population, they will too often give misleading or unrepresentative information on genetic variation.

Electrophoresis of proteins has been widely applied for the direct study of genetic variation in fish populations. The importance of electrophoresis to the study of fish genetics resides in the ability to directly estimate genetic relationships from its results, and also because variation of electrophoretically detectable genes is often correlated with variation of other genes. To the extent that such a correlation is widespread among fish species, electrophoretic variation can be a general estimator of genetic variation.

The "state of the art" is to use electrophoretic techniques to analyse genetic variation in natural populations because, among other things, electrophoretic variation is "taxonomically congruent" with morphological variation in interpreting phylogenetic and evolutionary relationships (Mickevich and Johnston, 1976). However the use of electrophoretic variation analysis requires some qualifications regarding aquaculture. At this time there is no direct evidence in the literature to indicate that either qualitative or quantitative allozymic variation is indicative of potential economic performance in such characters as food conversion rates, tolerance to temperature extremes, low dissolved oxygen tension, etc. Because of this, caution must be exercised in using the level of allozymic variation as the only criterion for choosing stocks for desirable physiological, nutritional, or other related production performances. There is some evidence of a correlation between electrophoretic variation, meristic variation, and developmental stability in nature (Soule, 1980). It is theoretically possible to develop the use of qualitative variation as predictors or indicators of quantitative performance in laboratory or hatchery populations but this involves complex breeding systems probably beyond the practical scope of most aquaculture breeders (Soller, et al., 1976). There are situations in which it may be desirable to electrophoretically estimate genomic variability and then to use these data as a base-line for comparing the genetic effects of a particular pattern of stock breeding or exploitation. For example, when exploitation of a species can be anticipated, base-line information on the genetic variation of pre-exploited stocks would be desirable. This would allow some direct assessment of the genetic consequences of exploitation by continuous monitoring of exploited stocks, or populations.

When re-introduction of a locally extinct population is contemplated, earlier base-line information might allow a closer matching of the introduced fish to the original population. Proper genetic matching would increase the likelihood of successful re-introduction. When base-line data is not available (the usual case), direct genetic assessment of the potential parental stocks for re-introduction allows an intelligent choice of stocks for introduction. Other things being equal, populations of maximum electrophoretic variation should be selected for introduction because this probably increases the likelihood of evolutionary adaptation to a novel environment.

A similar genetic monitoring of cultured species would be desirable to assess genetic changes that result from a particular culture scheme. It may be important to maximise outbreeding in a stock, in which case electrophoretic variation would be an important tool for monitoring the breeding programme.

There are certain dangers associated with the absence of genetic monitoring. In the south-western United States, a major breeding programme was undertaken some years ago in order to produce sterile males of the screw worm fly for introduction into the wild populations (Bush et al., 1976). During the breeding programme, a particular allele of the electrophoretically detectable gene dglycerophosphate dehydrogenase was accidently selected probably by natural selection in the culture population. This enzyme is important in flight metabolism. The properties of the selected allele militated against flight in the wild of the introduced, sterile males. Knowledge of the biochemical properties of the alleles and electrophoretic monitoring would have avoided this unfortunate situation.

We do not wish to imply or recommend that every exploited fish or every cultured stock be monitored in this way. Not only would this be expensive, but unless such studies are properly designed and controlled, the data are not likely to be very useful. We do, however, suggest that several carefully designed monitoring programmes be set up, and that these be coordinated to maximise their utility.

Population Genetic Structure in Relation to Exploitation and Extinction

Fishes probably surpass all vertebrate groups in their variety of social structures and kinds of life histories. It is not surprising therefore that some controversy has arisen regarding the significance of such variables as population structure, dispersal, and genetic drift, particularly with regard to genetic integrity of populations. At one extreme there are species like the American eel in which the adult population is spread out over thousands of kilometres, yet this species apparently verges on being a single random breeding (panmictic) population. At the other extreme there are hundreds of species which are territorial, which have demersal eggs with parental protection and which have very limited vagility. In species of this latter type with a localized and fragmented type of population structure, the neighbourhood size or the local population could be as low as 100, and there may be very limited gene flow between the local populations. As a consequence of this diversity, it is hazardous to generalise about demographic, georgaphic and genetic structural characteristics of fish. Each species must be examined as a unique case, even recognising the possibility for intra-specific variation in population structure.

A good knowledge of the population structure in the management of fisheries cannot be exaggerated, whether the purpose of the management is exploitation or preservation or both (as should often be the case). Only when stocks are properly defined can the fishery be managed optimally. For example, one might conceivably decide to artificially enhance a pink salmon fishery by introducing fry from a hatchery. But unless one knew that odd and even year pink salmon were genetically distinct populations, the resulting hybridization could cause significant genetic change and a decrease in fitness in both odd and even year stocks. Even very closely related sympatric species can have very different population structures, and it is dangerous to generalise from one to the next (Allendorf and Utter, 1979). In the rainbow trout in the Pacific northwest, the genetic data differentiate the populations into eastern and western groups, the major division coinciding with the crest of the Cascade Mountains. Many early workers had concluded that the principal basis for genetic separation of rainbow trout populations was anadromy and time of return to fresh water. Allendorf and Utter, however, emphasize that the taxonomic units based on electrophoretic data correspond to geographic groupings rather than to the above behavioural characteristics. They emphasized the importance of glacial events in dividing these populations historically.

In coho salmon, however, glacial events are not considered to have played a major role in the present subdivision of the populations in the Pacific northwest. The discontinuous distribution of (certain) transferring alleles among coho salmon populations cannot be directly explained on the basis of glacial events. In the chinook salmon the coastal populations appear to be genetically distinct from the inland populations in both Oregon and Washington for the fall-run salmon, but the line of geographical demarcation is different from that of rainbow trout. This information has significant management potential because it could allow the determination of the major areas of origin of ocean caught fish. Hence, among three closely related species in the same region, there is absolutely no correspondence in the geographic distribution of racially distinct populations, i.e., the geographic barriers that separate sub-populations in one salmonid species are not relevant in another. Management generalisations based on the distribution of populations in one species could prove disastrous if applied to other species.

Population structure is thus a useful guide for a priori priority ranking with regard to genetic resource preservation, at least with regard to the extinction potential of local populations. Geographic range alone is quite useful. The majority of species which reproduce in estuaries, river systems in the temperate zones, and in coastal pelagic zones are relatively widely distributed and fairly numerous. Species residing in tropical floodplain rivers, extreme environments such as shallow desert lakes or salt lakes may be far less numerous and typically have rather limited geographic ranges. Many such "local" species are relatively vulnerable to habitat disruption. Both from the standpoint of biological conservation and from that of genetic resource preservation. Species which are widely distributed and relatively numerous warrant less attention than do the species with very limited distribution and small population sizes. We do not mean to imply that genetic depletion is only a threat to local endemics.

The potential for genetic depletion can also be quite high for a single reproductive unit of a broadly distributed species. The obvious distinction between the two cases is that the extinction of the endemic cannot be reversed, whereas the recolonization of a reproductive habitat upon loss of the more broadly distributed species is possible. For example, the Japanese sardine Sardinops melanosticta following a massive population collapse and range contraction, has recolonized the Japan Sea from refugia on Japan's east coast. Another example is the successful, artificial reintroduction of Atlantic salmon in small coastal rivers in the northeastern U.S.A. In the latter case, the genetic characteristics of the replacement population should be a relevant concern. For example, failure to consider the particular ecological adaptations of replacement or enhancement of stock has caused severe management problems in bob-white quail (Clarke, 1954). In Scandinavia some introductions of Arctic charr and whitefish into lakes already populated by conspecifics has had deleterious consequences (Svardson, 1979). It is a moot point whether a naturally or artificially recolonized stock becomes as well adapted as the original stock. The point is to maximise the chances of a successful restocking or recolonization by taking into account all of the relevant genetic and ecological variables.

10

Post-process Handling

Delivery of the thermal process schedule must be strictly controlled to avoid under-processing spoilage; however, no matter how severe the process, product safety will be compromised if there is post-process leaker spoilage. There are several contributory factors leading to post-process leaker spoilage; these include the following:

- poor quality cooling water,
- poor post-process hygiene and sanitation. and
- container damage during handling and storage.

It is considered that even when can seam attributes comply with GMP guidelines for double seam formation, there is a shall number of cans which "breathe" or leak after seaming. Some estimates put this figure as high as 1% of all cans sealed.

The generally sound record of the fish canning industry suggests that, if this estimate is correct, only a fraction of those cans which leak ever spoil; this implies that either "micro-leakage" does not (necessarily) result in contamination, or not all contaminants are able to grow in the environment in the can. While this may be reassuring. there are no grounds for complacency. In 1978 and 1982 post-process leaker contamination by *C botulinum* type E was held responsible for the death of three people who contracted botulism after eating commercially canned salmon.

Chlorination and Cooling Water Quality

As product temperatures fall during cooling, there is a corresponding fall in the internal pressures in caps; and when the product temperature falls below the fill temperature a vacuum forms.

This means that the pressure differential across the ends of cans undergoing the final stages of pressure cooling, will favour the entry of cooling water into those cans in which there are seal imperfections. It is prudent, therefore. to accept the possibility of there being micro-leakage through the double seams of some cans (or glass closure seals, or the seals on laminated pouches) and that when this occurs cooling water will mix with sterile product.

On the few occasions that post-process leaker contamination does occur, it is important that the cooling water be of sound microbiological quality, for otherwise there is an unacceptably high probability of spoilage. It is because of the risks of post-process leaker spoilage that fish canners use sanitizing agents to control contamination levels in retort cooling water. Of those available, the most widely used are elemental chlorine and chlorine based compounds, however, other sanitizing agents include elemental iodine, iodine compounds and iodophors (a combination of iodine and a solubilizing compound which aids the controlled release of free iodine into the cooling water).

It cannot be assumed that sanitizers will be totally effective in eliminating contamination by viable vegetative bacteria and their spores; rather it is better to regard their action as being one which reduces the probability of survival to acceptable levels. Chlorine, for example is most effective against vegetative bacteria, less so against *Clostridium* spores and least of all against *Bacillus* spores. This is why the most likely contaminants in chlorinated cooling water are expected to be spores belonging to the genus *Bacillus.*

Chlorine may be added as gaseous chlorine (Cl2} which hydrolyses to form hydrochloric acid (HCl) and hypochlorous acid (HOCl, the agent which is responsible for the destruction of vegetative bacteria and spores). Hypochlorites may also be used for chlorination of cooling water, the most usual forms being as liquid sodium hypochlorite (NaOCl) or solid calcium hypochlorite (Ca[OCl]2). Irrespective of which form of chlorine is used. It is important to allow for the reactions that take place with inorganic and organic impurities in the water. When chlorine is added to commercial quality water, it first combines with these impurities (e.g.. minerals and nitrogen containing organic compounds) to form chloro-derivatives which lack the germicidal properties of free chlorine. As the dose is increased these are oxidized, at which point the chlorine demand of the water is said to be satisfied and the "break-point" reached. The chlorine residual remaining after break-point chlorination is called the "total residual chlorine". Total residual chlorine comprises the chloramines and chloro-nitrogen compounds (i.e. the "combined residual chlorine" which exists below the break point) plus "free available chlorine" (i.e. the free chlorine or loosely combined chloro-nitrogen compounds which exist above the break-point). Once the break-point has been reached the addition of more chlorine will lead to a proportional increase in the free available chlorine.

At the normal pH of cooling water free available chlorine is a more effective bactericide than combined residual chlorine. It is usual to dose cooling water so that free available chlorine remains detectable after a contact time of 20 min. Excessive chlorination of cannery cooling waters is wasteful and it also should be avoided because chlorine is corrosive to some metals. The lethal effect of chlorination increases at low pH (at levels where undissociated hypochlorous acid predominates), at high temperature and with high levels of free available chlorine.

There are practical constraints as to how low the pH can be, given that normal cannery cooling water is in the pH range of 6.5 - 8.5. Another constraint is that at high temperatures chlorine loses solubility and is driven off; elevated temperatures also make rapid cooling of cans difficult. High levels of organic matter increase chlorine demand, and, like inorganic impurities, they also protect bacterial contaminants.

Under GMP conditions it is sufficient to maintain residual free available chlorine levels of 2-4 mg/L after a 20 min contact time in order to be confident of holding total aerobic counts at less than 100 organisms/mL of cooling water. Free available chlorine should be still detectable in the cooling water at the completion of the cooling cycle. At all times records of free available chlorine levels should be maintained to provide confirmation that cooling water chlorination procedures were adequate.

Post-process Hygiene and Sanitation

It is known that when conveyors and can handling equipment down the line from the retort are unclean. they harbour high numbers of contaminants which can contribute to the incidence of post-process leaker spoilage. These basic hygiene problems can often be compounded because when cans pass from the retorts they are still warm, and this means that the plastisol lining compound in their ends will not have had sufficient time to "set up" and form a seal that is resilient to impact and deformation. Also at this stage the vacuum in the can will have partially developed, so that contaminants on and around the double seam are liable to be drawn into the container should the seal leak, even momentarily. Because of this, it is important to clean and regularly sanitize all those surfaces which come into contact with containers.

Conveyor guide rails, twist conveyors, transfer plates, elevators, push bars and accumulation tables should all be made of impervious materials which can be cleaned easily, thoroughly and regularly. In order that the containers are dry during post-process handling, it is good practice to include in the line, close to where the cans are unloaded from the retort, air blow-driers (or similar equipment).

These systems are preferable to the inappropriate plastic curtains which are all too frequently installed to drag over the surface of cans as they are conveyed underneath. The longer the cans remain wet, the greater the opportunity for post-process leaker contamination. For this reason containers should be dried as quickly as possible, so that exposure to wet post-retorting conveying and handling equipment is at a minimum. In line with GMP guidelines conveyors or equipment surfaces should be effectively cleaned every 24 h, as well as being disinfected during production, if they are wet while in use. Container drying may be accelerated by dipping the retort crates containing the cans into hot water containing a wetting agent. It is sufficient to submerge the crates for approximately 15 sec, and after they are removed from the bath they should be tilted to allow any adhering fluid to drain from the surface of the cans. If this procedure is adopted it is important that the dip tank be held at > 80 °C and that the water be changed regularly to avoid microbial build-up. Use of porous labeller pads and drive belts is discouraged, as these materials can provide an excellent environment for the accumulation and multiplication of microbial contaminants, particularly when they are wet, dirty and irregularly cleaned and sanitized.

When adhering to GMP guidelines and while implementing adequate post-process hygiene and sanitation procedures, manufacturers should comply with the following guidelines for bacterial counts on container contact surfaces and in the water entrapped in can double seams:

- Post-process conveying surfaces; pre-production, post-production and after sanitation procedures during production. Not greater than 500 org/mL.
- Water in double seams after cooling and handling. Not greater than 104 org/mL.

Final Operations

Container Damage During Handling and Storage

Poor: quality cooling water and/or inadequate hygiene and sanitation will increase the risks of post-process spoilage if containers are subjected to rough handling, particularly when this results in damage to the seal area. While unloading retort baskets extreme care should be taken to avoid mechanical damage to hermetic seals, and, because of the risk of contamination from operators, wet containers should never be unloaded manually. Conveyor systems in which line-pressure prevents easy removal of cans by hand need readjustment or re-design to improve flow. Severe can to can impact leading to damage at the end of twist conveyors is indicative of poor line design which provides an opportunity for contamination, because at the moment of collision the can compound is frequently still warm and soft, and the seams wet. Not all manufacturers find it appropriate to install semi-automatic or fully automatic equipment and so rely instead on manual handling to complete final operations. While this is often an attractive proposition, as it can be the cheapest. and most versatile mode of operation, it carries with it the heightened risk of post-process cross-contamination from operators and/or their protective clothing when metal cans, glass jars or laminated pouches are mis-handled. Therefore, wherever manual procedures are adopted, manufacturers must be sure that containers are dry and that operators handle them carefully.

Operators must be discouraged from using processed cans, whether packed in cartons or loose in "bright stacks", for other purposes; such as for bench supports, or for seats, or for racks on which to dry wet protective aprons and gloves. The reason for this concern is that in the fish canning plant there are assumed to be food poisoning spoilage organisms which could grow and render the product a threat to public health, if they are able to gain entry into the processed container and contaminate the contents. The potential danger of post-process contamination can be comprehended when it is recalled that the last three botulism outbreaks involving canned fishery products manufactured in the United States (i.e., tuna in 1963 and salmon in 1978 and 1982) are all alleged to have occurred because sterilized containers were contaminated with *C. botulinum* type E. The 1963 case was believed to be the result of faulty double seam formation in the canner's end; the 1978 outbreak was attributed to seam damage followed by corrosion leading to a small hole in the seaming panel;

and the 1982 outbreak was attributed to an indexing fault caused by a malfunction of a can reforming machine. In each incident spoilage through post-process leakage and contamination (by *C. botulinum* type E, or its spores) was implicated, rather than under-processing spoilage because the microorganism responsible was:

- relatively heat sensitive and therefore unable to survive even a marginal process;
- known to be widely distributed in the marine environment and therefore a possible contaminant of fish processing plants;
- non-proteolytic and therefore not a producer of the putrid odours which would normally deter consumers from eating the spoiled product.

The circumstances surrounding these outbreaks highlight the difficulties faced by all fish canners who, because of the origin of their major raw material, cannot avoid operating under conditions in which contamination by *C. botulinum* type E must be assumed to be the norm. Although a worst-case scenario such as this is extremely cautious, it confirms the need for extreme care when handling processed containers. Frequent jamming of conveyors and container handling equipment; indicates a need to re-appraise the machinery, the line design or the speed of operation because the potential risk arising from damage to the hermetic seal are untenable. Problems arising from poor handling are not confined to metal cans they do not fracture like glass, or puncture like retort pouches. However, because they are robust and because it is easy to overlook apparently superficial damage, cans are not always handled with appropriate care. Considering that in some countries post-process leaker contamination has been estimated to account for between 40 and 60% of canned food spoilage it is clear that the problems of post-process container damage ought not be underestimated.

Rate of Cooling

Cans should be rapidly cooled to 40 °C in retorts, otherwise they may remain at thermophilic incubation temperatures during labelling, packing into cartons, palletizing and storage. When rapid cooling in water is not possible, some manufacturers choose instead to air cool their product; should this option be favoured, care must be taken to ensure that there is unrestricted air circulation around the cans. A further problem associated with inadequate cooling is stackburn which results from the over-cooking that occurs when product is stored while still hot.

Temperature of Storage

Selection of storage temperatures for canned produce may be critical for those products containing thermophilic spore-forming survivors. Target Fo values are generally more than sufficient to kill mesophilic spore-forming contaminants provided that raw materials are of reasonable microbiological quality. However, because ambient temperatures in warm climatic zones often encourage the growth and multiplication of thermophiles, processes must be either sufficient to reduce, even these extremely heat resistant bacteria, to a satisfactorily low level (e.g., < 1 in 10^2), or storage must be at temperatures unfavourable for their growth. In addition to the concerns about storage temperatures, it is recommended that canned fishery products be stored under conditions which avoid sweating caused by extreme temperature fluctuations, as this phenomenum will encourage external rusting of the containers, particularly in areas of high humidity. These conditions are to be avoided also where containers are packed in retail cartons or outer shipper cartons as these will absorb moisture and may even collapse in the warehouse.

Canning Processes

In this Chapter are summarized the stages in the production for each of the main commercially produced canned fishery items. Where typical retorting schedules are given, they are intended for guidance and should not be adopted without first having their adequacy confirmed through heat penetration trials conducted under commercial operating conditions, or in laboratories equipped to conduct these determinations. In addition to the requirement for product safety and shelf stability. canned fish are expected to have sensory properties which are characteristic of the species, and the product must be free of objectionable odours, taints or visual defects. Major product compositional and quality requirements are specified in the set of Codex Standards for Fish and Fishery Products (CAC/ VOL. V -Ed. 1.1981 1/), which include specifications for the following canned products:

Salmon, Canned Pacific	Codex Standard	3 - 1981
Shrimps or Prawns, Canned	Codex Standard	37 - 1981
Tuna and Bonito. Canned in Water or Oil	Codex Standard	70 - 1981
Crab meat, Canned	Codex Standard	90 - 1981
Sardines and Sardine type Products Canned	Codex Standard	94 - 1981
Mackerel and Jack mackerel Canned	Codex Standard	119 - 1981

Sardine and Sardine-like Fish

Sardines are usually canned by one of two methods; the first is inferred to as the traditional Mediterranean method (so named because of its origin, although nowadays similar technology has been adopted elsewhere and is generally described as the "raw pack method" and the second is a method incorporating a hot smoking step, rather than in can pre-cooking. The latter method is commonly practised in Western European countries.

Traditional Mediterranean Method

Either fresh or frozen sardines can be used to produce a good quality canned product provided that the preliminary handling conditions have protected the fish from excessive deterioration during transport and storage. The sequence in which preliminary operations are carried out varies from processor to processor, and may reflect such things as the complexities of the line, the speed of production, the degree of automation, the availability of labour and the Source and type of raw materials. One sequence for a Mediterranean style canning line is as follows: the sardines are weighed and washed and then, brined (by immersion in a saturated solution for up to 15 minutes, depending on size and fat content), graded, nobbed and packed. In an alternative sequence nobbing precedes brining so that the order of the pre-treatment operations becomes; weighing, washing, nobbing, brining (in batches, or continuously in screw conveyors, to a final salt content of between 1 and 2%) packing and washing - although a recent modification of this procedure includes direct addition of salt to filled cans, which means the brining step can be eliminated.

Sardines are fed automatically or manually to the nobbing machines in which the heads, viscera and tails are removed. The machines are set to cut the fish to standard lengths, or into cross-out pieces, so that pack uniformity is achieved. Machines are available which complete the traditional nobbing operations and then pack the fish into cans automatically. however in many cases packing remains a manual operation.

Pre-cooking of the sardines in filled cans is carried out in automatic steam cookers. The first stage is a steamer, operating at around 95 °C, through which the cans pass while held inverted on perforated conveyors to allow simultaneous entry of the steam and drainage of condensate and oil exuded from the flesh. In some pre-cookers the cans are steamed in the upright position but inverted and drained before passing to the second stage. The final phase of pre-cooking is a drying process taking place at around 130 °C. As an alternative pre-cooking method, some canners fry their sardines, but this is generally more expensive.

Cans containing pre-cooked fish pass to a liquid filling station where one of either brine, water, edible oil, sauce or marinade is added manually or automatically. For those products which have not .been brined, the salt is added in solid form prior to the addition of the liquid medium, or it can be blended with the liquid. The cans are then transferred to can sealing machines for double seaming with pre-coded can ends. While adding the liquid, there is usually some overfill which can be recovered. When the outside of the cans are contaminated with oil and/or fish remnants they should be washed in water and detergent before they pass to the retort for sterilization.

It is preferable that can washing be completed prior to, rather than after, sterilization so that the risks of post-process contamination are reduced. If post-process washing cannot be avoided, it is essential that it be carried out in hot water of sound microbiological quality and preferably in which there has been included a surface active agent to assist can drying -at all times extreme care should be taken to ensure that processed cans are not manually handled while still wet. Sterilization is usually in batch retorts in which the heating medium is either pure saturated steam, hot water or recirculated hot water which is pumped over the cans.

Because of the large surface area, and therefore flexibility. Of the traditional club, dingley and hansa style sardine cans the ends are prone to distort as a result of the internal pressure generated during processing.

This can be compensated for by processing the cans under an over-pressure; but more importantly cans should be pressure-cooled, at least until the internal can temperatures have fallen. Retorting temperatures and times are selected to suit the desired textural properties and the target Fo value of the process. As a o general guide 1/4 club and dingley cans are processed for between 45 and 60 min at around 115.6 °C; although some processors choose to process at 112 °C, because they find the bone softening at these temperatures to be preferable to that achieved by a shorter process at a higher temperature.

After cooling the cans are dried in air, packed in individual cartons and then into master cartons. Before release the finished product is held to ripen in order to develop the characteristic flavour and textural properties.

Norwegian Method

The major difference between the Norwegian and Mediterranean methods of canned sardine manufacture is that with the former the fish are not eviscerated and are usually hot smoked, whereas the Mediterranean method includes evisceration and pre-cooking. With the Norwegian method evisceration is unnecessary because the catch is held alive for at least 48 hours in nets prior to landing (the holding process is known as thronging) during which time the fish digest their feed and thereby minimize the enzymic activity which if left unchecked would lead to belly-burst, while the smoking process replaces flash pre-cooking. After thronging the fish are transported fresh to the factory for immediate use or for frozen storage. The following description is typical for the traditional Norwegian method of canned sardine manufacture.

If frozen raw material is used the blocks are thawed under running fresh water or sea water. When thawed with sea water. It is often not necessary to brine the fish; however when using fresh water. Or when fresh fish are used. The sardines are flumed in a brine solution which washes them, removes scales and enables the fish to absorb from 1 to 2% salt.

The fish are automatically size graded and passed into threading machines where they are fed through a series of parallel' plastic pipes out of which they emerge. One fish at a time. In rows. Metal rods (spits) are threaded through the eyes of the sardines. A row at a time. The spits are hung on frames and the frames are then stacked on trolleys and transported to smoking ovens.

The drying and smoking process takes place in the smoking oven. The temperature of which is set to suit the size and fat content of the fish; typical inlet temperature is between 40° and 60 °C, while normal outlet temperature is between 120° and 140 °C. The total drying and smoking process takes approximately one hour. The hot air for drying is derived from steam heated heat exchangers, while the smoke is generated by burning oak or other hardwood chips.

The removal of moisture in the smoking oven prevents the release of excess water during retorting. And the addition of smoke gives the sardines their characteristic flavour. In some instances canners use artificially flavoured oil to impart the "smoked" taste. However. When this technique is used there must be an accompanying declaration on the label; in these circumstances the other stages of the process are similar to those described in the Mediterranean style production line.

In the Norwegian hot smoking process the fish are smoked, while hanging on the spits, and then passed to rotating knives where their heads are removed with a cut directly under the gillbone. The bodies fall into trays below and are transferred to the filling floor for hand packing into cans. Filled cans are automatically conveyed to an in-line oil filler from which they then pass to a can double seamer where coded ends are applied. Sealed cans are transferred, via a flaming channel, to retort baskets sitting immersed in a tank containing hot water and detergent in which contaminating oil is washed from the outside surface of the containers.

The cans are processed in counterbalanced retorts in which an overpressure is necessary to prevent deformation (caused by the high internal pressure generated in the cans during the thermal process) of the relatively large and flexible ends.

Typically for 1/4 dingley cans the total pressure in the system is approximately 122 kPa (18 psig). This means that when processing at 112 °C, the pressure due to the steam vapour pressure will be around 52 kPa (7.5 psig), while that due to the air overpressure will be approximately 73 kPa (10.5 psig). Retorting temperatures and times vary, but generally for 1/4 dingley cans, the process is for 60 min at 112 °C which is sufficient to deliver a target Fo of > 6 min. The retorts use conventional steam heating or, alternatively, the heating medium can be water which is pre-heated in overhead vessels and then dropped into the retorts below, where it is brought to operating temperature with steam under pressure.

After sterilization, cans are dried in hot air and passed to automatic or manual case-packing for packing into cartons. As with Mediterranean style sardines, it is necessary to hold the finished product to allow it to "ripen" and develop fully its characteristic sensory properties.

Tuna and Tuna-like Fish

There are several styles of canned tuna described in Codex Standard No 70 (referred to at the beginning of this Chapter); however, apart from minor handling differences arising from variation in the size of the species and the pieces, the relative proportions of light and dark meats, and the styles of liquid fillings, the stages in the canning processes are substantially the same.

The pre-treatment stages include thawing in running water, heading and evisceration of the smaller species (which are usually frozen whole). Larger fishes are headed and gutted on board prior to freezing. Once thawing is complete, or when fresh chilled fish are used, the fish are cut into vertical pieces, or horizontally into loins, and washed and placed on metal trays which are transferred on racks into the atmospheric steam pre-cookers. Pre-cooking is carried out in steam at between 100° and 105 °C for as little as one hour for small species, or over eight hours for large specimens.

The temperature and time combinations of pre-cooking are often regarded by canners as being critical to their overall yields; generally, the common aim is to raise backbone temperatures to between 60° and 85 °C, after which the portions are removed from the cooker and allowed to air cool, often overnight.

In climates where ambient temperatures are around 30 °C, or more, it may be necessary to assist cooling by placing the fish into chilled storage, so that the flesh will not be held for too long in conditions favourable to contamination or microbial activity. Cooling can also be achieved by water spray in order to hasten the process. After cooling the flesh firms, which makes the subsequent cleaning and picking operations easier for the operators.

If not already done so, the head, tail and fins are removed; the skin is scraped from the flesh surface and the white and dark meat portions picked from the frames and segregated. The edible portions are selected for solid, chunk, flake or grated (shredded) style packs and then transferred to filling areas. In many of those countries where labour costs are relatively low, packing is a manual operation; however, there are machines which perform these tasks fully automatically for all styles of packs.

The filled cans are transferred to brine or oil fillers, or in some cases they first have dry salt added, after which the water, oil or sauce is added. Cans then pass to the can seaming machine where they are closed under vacuum by coded can ends attached in a double seamer. The hermetically sealed cans are manually or automatically loaded into the retort baskets of manually operated batch retorts, or they may be directly conveyed into crateless retorts or hydrostatic retorts for sterilization.

The temperatures and times selected for retorting depend on the container size, the pack weight, filling temperature and the pack style. Generally, while operating under GMP conditions it is sufficient to process to Fo values of around 10 to 15 min; however this there is evidence that some canners select unnecessarily severe conditions which deliver Fo values in excess of 30 min. Apart from being wasteful of time and energy, such severe processes adversely affect the sensory characteristics of the product. As a guide to selecting temperatures and times for processing, a summary of the conditions used commercially for a variety of can sizes is shown in.

Table: *Typical retorting conditions for tuna processed at 115.6° and 121.1 °C in a variety of can sizes*

Can dimensions		***Retorting time***	
Diam. (mm)	***Height(mm)***	***115.6 °C(min)***	***121.1 °C(min)***
66	40	65	40
84	46.5	75	55
99	68.5	100	85
154	118.5	230	190

After thermal processing cans are cooled (preferably under pressure, although some manufacturers pressure cool only their larger cans because they are the ones most likely to peak), dried in air or with the assistance of air blowers, and held in "bright-stacks" prior to labelling and packing, or labelled and packed directly off the line.

Salmon and Salmon-like Fish

Fresh or frozen salmon are transported from storage and graded before passing to the iron chink machine for dressing, an operation which automatically removes the head and tail, splits the belly, and removes the viscera and fins. After butchering, the fish are transferred to the sliming table for the final removal of flesh and blood remnants and for washing.

The fish are cut to size for automatic or manual filling into pre-washed cans. Filled cans have salt added and then pass, via a check weighing machine, to a can seamer for vacuum sealing (using coded can ends). Between filling and seaming, cans pass to a "patching" station where operators check for traces of skin, bones or meat lying across the flange of the can; while making adjustments to pack fill weights if required.

Sealed cans are washed, packed into baskets and then loaded into retorts for processing. There should be no more than one hour's delay between container filling and the commencement of the thermal process as longer delays may lead to pre-process (incipient) spoilage. Cooled dry cans are either bright-stacked or labelled directly and then transferred to warehouses for storage.

Crustacea

Crab

The preliminary stages of cooking and picking of crab meat are simple operations, often suited to small scale operators working in relatively unsophisticated conditions, but who are close to the supply of the raw material. It is important that the crabs should be handled under conditions which limit the opportunity for degradative enzymic action leading to the deterioration of their fresh flavour. This means that, ideally, the crustacea should be held wet and cool as soon as they are caught; and that preferably they should be cooked alive, or as soon as possible after death. Cooking can be either in boiling water, or in retorts using steam under pressure. If the cooked meat is not to be butchered and picked immediately, it should be refrigerated or iced.

After cooking, the crabs are washed to remove sand and the scum that adheres to the shell after cooking. The claws and legs are removed from the body which is then eviscerated. It is important that the butchering operation be carried out separately from picking, and that at all times the opportunities for cross contamination are avoided by implementation of hygienic handling practices. Shell particles are removed and then the legs, claws and washed bodies are held ready for picking.

During the picking operation fragments of meat are drawn from the claws, legs and body either by hand or in some cases using automatic equipment. In the latter cased, the shell is crushed and the flesh separated by brine flotation which allows the shells to sink and the meat to float. A disadvantage of mechanical picking is that the flesh tends to become shredded and the structure of lumps is damaged. Pickers separate the flesh from the shell, and segregate it into grades with the best prices being paid for flesh in which the structure of the piece is retained. If there is a delay prior to canning. the picked flesh is stored under refrigeration so as to avoid contamination. Because even though the meat has been cooked, it is nevertheless vulnerable to microbial deterioration, particularly in warm climates.

On receipt at the cannery (if the picked flesh has come from elsewhere), or prior to further treatment (if the crabs have been picked on site), the meat should be inspected for uniformity of grade, acceptable odour (off-odours indicating deterioration), satisfactory colour and the absence of shell and other contaminants. Once passed by inspectors, the flesh may be blanched in boiling water- this operation is optional, and is more usual when pre-cooking operations have not totally cooked the flesh. Blanching firms the flesh and protects yields, which otherwise are found to be depressed if pre-cooking is incomplete. Some manufacturers include metabisulphite in the blanching brine, and/or soak the crab prior to pre-cooking -both treatments are used to prevent discolouration. After blanching. The hot meat is cooled in potable water, or in air.

As detailed in the Codex Standard, various pack styles are available. The flesh is packed in lacquered cans to prevent sulphur staining and in some cases parchment is also used. The filled cans are topped up with brine containing 2-3% NaCl, and in some circumstances a 0.1-0.5% citric acid solution to prevent discolouration. Cans are then vacuum sealed and retorted. Typical processing conditions are shown in. After processing and cooling in chlorinated water, cans are either bright stacked, or labelled and then transferred to the warehouse for storage. Some canned crab tends to discolour and form a blue/black or grey/black pigment. This reaction has been discussed by Howgate (1984), who pointed out that the mechanism is not clearly understood, partially because the phenomenum appears to be species related. Howgate summarizes three current explanations for this discolouration, which have given rise to three different solutions to the problem:

a. The blueing is due to the presence in the flesh of copper; the solution is to include in the brine a metal chelating agent such as citric acid, or ethylene diamine tetraacetic acid (EDTA).
b. The grey discolouration results from a variation of the well known Maillard browning reaction which occurs between sugars and amino acids at high temperatures; a partial solution is to lower retort temperatures and increase processing time (e.g., process at. 115.6°C rather than at 121.1°C), and/or to include sulphur dioxide in the brine.
c. The discolouration is the result of melanin formation, derived from an enzymically related oxidation of tryosine; the solution is to expose the flesh to a sodium metabisulphite treatment as a dip, or include it as an additive in the blanch water.

While processors may find any or all of these solutions acceptable as a means of controlling or eliminating discolouration, they must first assure themselves that regulators in importing countries permit the inclusion of the additives which overcome the problem.

Shrimp

Raw shrimp should be received refrigerated, or well iced, to limit enzymic and microbial action which, if unchecked, leads to lost of quality. Some canneries receive their shrimp peeled and cooked, however, even under these conditions the need for adequate cool storage cannot be neglected. The shrimp are inspected on receipt at the cannery, and then washed to remove adhering dirt and ice. Peeling can be either manual or by machine. In an example of the latter, the shrimp are size graded, and the head and shell removed by a combination of gentle pressure and a controlled rolling action. The final pre-treatment is to devein the shrimp after which they are washed and reinspected.

Cleaned and shelled shrimp are pre-cooked in hot brine, or steam. The choice of salt concentration and cooking time varies from processor to processor. Generally, salt concentration will range from 3 to 13% NaCl, and pre-cooking time will range from 2 to 10 min; the conditions chosen will be affected by shrimp size, the temperature of the solution and whether the shrimp are to be for a wet pack (i.e., packed with brine) or for a dry pack. Steam pre-cooking is usually carried out at around 95° to 100 °C for 8 to 10 min, depending on shrimp size. The shrimp are then cooled, dried, inspected and size graded, prior to hand packing in cans which have been lacquered to resist the formation of unsightly black sulphide stains. Filled cans are topped up with hot brine (to which some processors add citric acid to reduce discolouration) and then sealed. Vacuum in dry pack cans is achieved by either exhausting the cans prior to sealing, or, alternatively, by sealing the cans in a mechanical vacuum closing machine.

The difference in the processing times required for wet and dry packs in the same sized containers, arises because of the convection currents in the brine pack increasing the rate of heat transfer to the SHP of these containers. At the completion of the thermal process, the cans are cooled with chlorinated water and removed from the retorts for either bright stacking or direct labelling and packing after which they are transferred to the warehouse for storage.

Molluscs

Abalone

The best quality canned abalone is manufactured from the fresh product, although some canners may use frozen stocks, however, when they do, their yields are decreased and the texture of the finished product tends to be too soft. Fresh abalone is received chilled and is then shucked by hand before the meat is transferred to washing tanks for the removal of the pigment from around the lip. The cleaning is achieved by immersing the abalone in warm (35 °- 40 °C) water for approximately 30 min, during which time the flesh is gently abraded by rotating the tank holding the brine or by stirring the brine with paddles. Some canners use proteolytic enzymes to assist removal of. The pigment, in which case it is necessary to arrest enzymic activity by dipping the molluscs in a solution of hydrogen peroxide. Cleaning is completed by gently scrubbing the flesh with nail brushes or abrasive pads.

Good quality canned abalone has a creamy/yellow colour, however under some circumstances, not always clearly understood, there is a blue surface discolouration. It believed that the mechanism for this action is related to the formation of a metallic complex, which explains why it can be controlled by the addition of chelating agents such as citric acid and/or EDTA; it can also be controlled by the addition of metabisulphite. These additives may be included at a number of stages in the pre-treatment; such as in the cleaning brine, in the blanch water, or in a dip. They may be added to the canning liquor, provided that the country in which the product is to be sold does not prohibit their inclusion.

After cleaning, the abalone are trimmed (to remove the viscera and gonads) and then blanched for 5 min at 70°C, before packing into lacquered 74 x 118.5 mm cans. Most manufacturers pack three to four whole abalone per can (individual abalone weights can range from approximately 90 to 180 g), and make up to minimum pack weight (i.e. , usually 50% of net weight) with portions. The cans are topped up with brine containing approximately 2% NaCl and vacuum sealed with coded can ends. If hot brine is used, it may not be necessary to vacuum close the cans, however, under these circumstances it is important that the cans be pressure cooled in the retort.

During thermal processing there is a textural inversion associated with the softening of the pedal sole and the toughening of the myofibrillar proteins at the base of the adductor muscle. There is also a weight loss, which can account for reductions in yield of between 12 and 30% of fill-in weight under extreme conditions (e.g., with a severe thermal process combined with the use of stale or frozen stock). Given the high selling price of this commodity, manufacturers are therefore keen to avoid overprocessing, without compromising the safety of the product. This means that despite the unavoidable weight losses caused by retorting, canners must still be sure that their minimum target Fo values are>2.8 min.

Fish Pastes and Spreads

Fish pastes and spreads are manufactured from by-products or from under-utilized species, for instance those fish which are generally too small for other purposes. They usually require mincing and/or blending with other ingredients such as salt. Sauces, spices, fat, emulsifying agents and thickening agents. Often they are sufficiently viscous to be filled as liquids or pastes in automatic filling machines as well as in manually operated devices. In some cases a blended mixture is formed into fish balls or fish cakes which are hand filled and topped up with brine or sauce. The scope for development of these items is large, and they have particular appeal as a relatively cheap, yet nutritious fishery product.

Generally, the products are packed into lacquered cans or glass jars, and sealed with coded ends and caps, respectively. Processing of cans is in conventional retorts, whereas glass jars, retort trays and cylindrical shaped flexible laminate packs require thermal processing in counterbalanced retorts. Some canners produce fish paste and seafood spreads in retortable aluminium trays and pouches. Because of the flat profile of these packages, heat penetration to the SHP of the container is rapid (relatively), which means that heat sensitive products (e.g., scallop and lobster pastes) can be processed without excessive loss of flavour and colour.

Processing temperatures and times depend upon the dimensions of the container and the nature of heat transfer to the SHP of the pack -for pastes this is largely by conduction, while for fish balls in brine it is by a mixture of convection and conduction. Post-process handling procedures are similar to those for other heat sterilized fishery products.

Equipment for Fish Canning

For detailed descriptions of machinery used in production of the major , commercially canned fishery products, and for accompanying flowsheets, reference should be made to the FAO Fisheries Circular No 784. Planning and Engineering Data. 2. Fish Canning (1985), in which can be found also, examples of plant layouts for the major species canned. This Chapter contains a description of the processing equipment specifically used for the production of canned sardines and tuna (following the procedures outlined in Chapter 4), together with a description of the thermal processing equipment which is basic to most fish canning operations.

Machines for Canning Sardine

Grading Machines

Grading machines are used to sort sardine and sardine-like fish into regular sizes. Machines are available which, in a single pass, segregate the fish into four different grades, with thicknesses ranging from between 5 mm and 33 mm.

The fish pass, tail first, down inclined oscillating tracks which are separated by gradually widening gaps. When the gap between the tracks becomes greater than the thickness of the body, the fish fall through to belts below, from where they are segregated into storage bins or passed onto conveyors for further processing. The machines are fitted with water sprays which simultaneously wash the fish as they pass down the tracks.

Shown in Figure 24 is an example of a grading machine supplied by the Baader Company of West Germany. In the figure can be seen the feeding mechanism which deposits the fish onto the tracks down which the fish move while being graded for size. The Baader machine shown is designed to grade herring, mackerel, sprat and capelin.

Nobbing Machines

There is a range of nobbing machines available for the removal of heads, tails and viscera of sardines and sardine-like fish; and there are also machines which automatically pack the nobbed fish into cans. Fish may be fed to the nobbing machines manually, by between three and five operators; however, there are also machines in which one supervisor can manage an automatic feeding operation. An example of a Baader feeding machine designed to handle herrings, sardines, sprats and similar fish. The fish are raised on an elevator and passed to five feed channels for delivery to the nobbing machine (or other piece of processing equipment).

Once placed in the nobbing machine the fish are fed to cutting knives which shear the head from the body without cutting the throat. The head is then pulled away from the body, after which the rotating action of tapered fluted rollers remove the viscera. Shown in Figure 26 is a simplified sequential sketch of the fish as they pass through a Baader nobbing machine. The machines can be set to leave the tails on the fish, or alternatively, the tails can be removed and the body can be cut to standard lengths (in one or several pieces) to suit the size of the Cans. Two pack styles available for sardines and other similar fish. Machines are available to handle fish ranging in length from approximately 10 to 45 cm, at a rate of 150 to 450 fish per minute (depending on fish size).

In automatic nobbing and packing machines, fish are placed in moulded pockets (to suit the pack style) in which they are conveyed, in can lots, under rotating blades for the removal of the heads and tails. The fish bodies are then eviscerated by a suction process, after which they are automatically transferred to cans. In many traditional canneries the nobbing process is automatic but cans are still packed by hand on conveyors.

Flash Cookers

Sardines are cooked and dried in flash cookers in open cans which are automatically transported through continuous machines. There are at least two systems available, however in each, the mode of operation is similar. Filled cans are automatically fed into a steam heating section where the sardines are cooked.

For machines in which pre-cooking takes place while the cans are in the upright position, the filled containers are inverted to allow draining; however, when cans are inverted during pre-cooking, draining is continuous.

At the completion of draining the sardines are dried, and the cans then proceed to the automatic discharge unit. In Figure 28 is a simplified line drawing showing the side elevation of a pre-cooking machine (supplied by Trio) in which the fish are pre-cooked and then the cans drained for approximately three minutes, prior to being returned to the upright position for drying.

The speed of the machines may be altered to suit the load and container size, typically the machines process in excess of 10 000 filled cans per hour.

Smoking Ovens

In the manufacture of Norwegian style sardines, the fish are smoked either in a batch or a continuous system. The units consist of the drying chamber into which hot air (at around 40 °C) is drawn. And a smoking section. Smoking ovens may be either simple batch operations into which are placed trolleys containing the smoking frames loaded with fish; or they may be automatic systems which continuously draw the fish on their frames through the drying and smoking chamber. Equipment used for the remainder of the sardine canning process is similar to that described under the heading "General fish processing machinery".

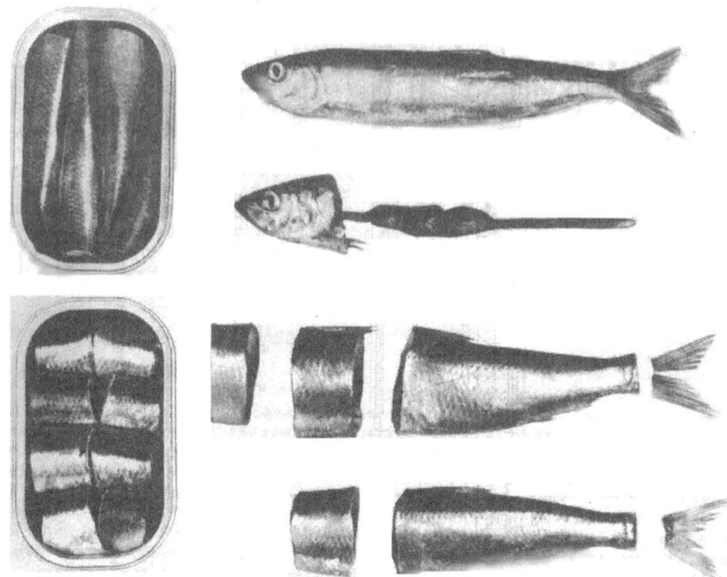

Figure: *Examples of pack style available with .automatic nobbing machine (Photograph courtesy of Baader)*

Machines for Canning Tuna

Pre-cookers

The most common pre-cookers are live-steam cookers. Fitted with condensate drains. Vents and safety valves. The pre-cookers operate on a batch system. With doors at each end (so that fish may be rolled in and out on a flow-through basis). The fish are loaded into galvanized iron baskets. And the baskets are placed on racks which are rolled into the cookers for steaming.

Other preparatory stages taking place before filling are completed manually. And in many canneries. Filling is also a manual operation. There are. However. Fully automatic filling machines suitable for packing tuna in all pack styles in round and oval cans.

Filling Machines

Machines are available for filling chunk and grated (shredded) tuna which operate at speeds of between 80 and 350 cpm with cans ranging from 112 to 445 9 (approx.). There are a number of manufacturers with various operating procedures, but one manufacturer (Carruthers Equipment Co., USA) has several machines for automatic tuna filling. In one machine (the Pack-Former) fish is discharged into filler bowls from where it is transferred into a series of piston pockets positioned around the circumference of the machine. As the filling heads complete a revolution, the fish is compressed into a cylindrically shaped slug in the pocket, in which form it is pushed out the bottom of the piston and is trimmed to the correct length, so that the weight of the pack in each can is controlled.

The fish is then fed into the can which has been located below. A machine (a Carruthers Nu-Pak) operates on a similar principle, at speeds ranging from 200 to 600 cpm, with 225 g cans (and smaller).

Solid style tuna loins are packed fully automatically by a machine (a Carruthers Pak.-Shaper) which handles cans ranging in size from 112 g to 1.8 kg (approx.) at speeds from 30 to 130 cpm. The machines are fed with solid loins which are transferred to a forming hoop in which the flesh is molded into the desired shape and then cut off cleanly to produce segments of the required length (and therefore weight).

Equipment used for the remainder of the tuna canning process is described in the following section.

General Fish Processing Machinery

Brining Machines

Brining machines are sometimes coupled with washing machines, so that the two operations occur simultaneously. In continuous applications, the machine is usually a rotating perforated drum partially immersed in a brine bath and through which the fish pass at a predetermined rate.

In less sophisticated operations. Brining can be a batch process in which the fish are loaded into perforated drums which rotate and, because of the tumbling action, gently transport the fish through the salt solution. Whether using automatic, semi-automatic or batch equipment, it is important that, the salt concentration be maintained at the desired level -this means that periodically the effects of gradual dilution must be monitored and salt added. The material used for construction of the equipment must resist the corrosive effects of the salt.

Exhaust boxes

The exhaust box is used to heat the contents of cans, so that they may be sealed hot, thus ensuring that, after cooling, a vacuum has formed in the container. Exhausting also drives entrapped air from the pack. Exhaust boxes may i take many shapes and forms, depending on the requirements of the cannery; basically they consist of a tunnel through which the open and filled cans pass while being exposed to atmospheric steam. They require a feed and a discharge mechanism, and a conveying system for transporting the cans from one end to the other. Recent models are frequently constructed with stainless steel, however many canneries still find painted mild steel systems adequate.

Sealing Machines

When selecting can sealing machines. Fish canners must consider the following factors:

- the size and shape of the container,
- the anticipated speed and volume of production,
- the level of skill required to maintain the machine in good working order.
- the cost and availability of spare parts, and
- the ease of "changeover" when the machine settings have to be converted to accommodate cans of more than one shape and/ or size.

In order to cater for the diverse requirements of fish canners, there is a wide range of machines from which manufacturers can choose a model to suit their operations. Since many sealing machines have features in common, the following is confined to a general description of the major categories which are readily available.

The simplest of machines are required by those packers who run their lines at speeds of from 8 to 25 cpm using hand operated or semi-automatic single-head equipment with motorized drives. For those with a low output (i.e., < 20 cpm), hand operated models are ideal - as with seasonal production or in those plants which are required to prepare test packs.

Single head seaming machines may be fitted with steam-flow closing or mechanical evacuation apparatus as a replacement for, or as an adjunct to, hot filling or exhausting. When mechanical vacuum closing is required the operator places the container (with the can end sitting in place on top of the can) in a chamber, which is then closed and evacuated by opening a line leading to a vacuum pump. When the desired vacuum is obtained in the chamber, the sealing operation is initiated by depressing a foot pedal which lifts the can up to the chuck on the sealing head and into position for double seam rolling. The first and second action rollers are sequentially brought into action while the can is rotated by the spinning seaming head. At the completion of the seaming operation the sealing chamber is opened to the atmosphere and the hermetically sealed container is removed. Machines of the type described can frequently have the facility for steam flow closing, in which case steam is injected across the headspace of the container (while it is positioned in the sealing chamber) immediately prior to double seaming.

Fully automatic in-line single-head steam flow closing machines which operate in the range of 70-90 cpm are available; while for canneries operating at higher speeds there is a variety of multiple-head machines from which to choose. Of the latter, three, four and six spindle machines are common and can be selected to cover seaming speeds of from 200 to 600 cpm, depending on can sizes and production capacity. Machines for sealing glass containers generally do not operate at the speeds of can closing equipment, however, they can be fitted for steam-flow closing or mechanical evacuation. Fully automatic steam flow closing machines are available to apply caps at around 400 to 500 cpm (depending on container size), while semi- automatic machines can be operated at around 15 cpm. As with cans, vacuums in glass jars may be also obtained by hot filling, or by addition of hot brine, or by exhausting.

Laminated packaging materials are sealed by the fusion of the two facing layers of the innermost ply. The material is heated while clamped between jaws of the sealing machine for sufficient time for the two layers (usually polyethylene or polypropylene) to fuse and form an hermetic seal. One of the greatest difficulties faced by users of laminated packaging materials is that of ensuring effective seal formation.

Under all circumstances the sealing surface must be clean and free of particulate matter, which can present difficulties when packing fish products, as it is not always possible to prevent flakes of flesh from contaminating the sealing surfaces. The solution to the problem is to clean the seal area before passing the package to the sealing machine, however, this further retards what is in many cases an already slow sealing operation.

Retorting Systems

For a detailed description of recommended retorts and retort fittings reference should be made to the following publications:

- *FAO/WHO* 1977: Codex Alimentarius Commission, Recommended international code of practice for canned fish. Rome, FAO/

WHO, Joint FAO/WHO Food Standards Programme, CAC/RCP 10-1976: 42 p. Issued also in French and Arabic

- *FAO/WHO 1983:* Codex Alimentarius Commission. Recommended international code of practice for low-acid and acidified low-acid canned foods. Rome, FAO/WHO. Joint FAO/WHO Food Standards Programme, CAC/RCP 23-1979: 50 p.

The main types of retorts used in the manufacture of low-acid canned foods include the following:

a. *Batch retorts heated with saturated steam:* These may be either vertical or horizontal and are by far the most common retorts used by fish canners. Simplified drawings of these types of retorts are shown in Figures 29 and 30; in Figure 31 is shown a less frequently used batch system for processing cans in saturated steam. The latter system is referred to as a crates. Brief descriptions of these systems are found in sections 3.6.1, 5.3.5 and 5.3.6.

b. *Batch retorts heated with water under pressure:* These retorts are vertical or horizontal and are most frequently used for processing glass containers which cannot be processed in pure steam because of the risks of thermal shock breakage. They are also widely used for sterilization of products packed in aluminium cans with score-line easy open ends. Simplified drawings of these types of retorts are shown in Figures 32 and 33; operational guidelines are given in section 3.6.2 and features of the system are described in section 5.3.7.

c. *Continuous retorts (other than hydrostatic retorts):* Containers are passed through a mechanical inlet port into a pressurized chamber containing steam where they are processed before passing through an outlet port and. Depending on the make of the retort. Into either another pressurized shell. Or an open water reservoir. For cooling. The motion of the cans through the retort causes some forced agitation which aids the rate of heat transfer to the SHP of the container.

d. *Hydrostatic retorts:* A simplified drawing of this type of retort is shown in Figure 34 and the system is described in section 5.3.8.

e. *Retorts heated by a mixture of steam and air:* The containers are processed under pressure in a system which relies on forced circulation (by a fan or a blower) for the continuous mixing of the steam with the air. Inadequate mixing can result in the formation of cold spots which could lead to under-processing spoilage. As with water filled retorts. This system is suitable for retortable pouches which require a counterbalancing overpressure to prevent their rupture.

There is a comparatively rarely used retorting system whereby sterilization is achieved by directly heating cans with flames from gas burners positioned underneath containers which spin past on guide rails. This system is suitable for packs which contain a high proportion of liquid, thus permitting rapid transfer of heat by convection, but it is not used commercially in fish canning operations. The most frequently used style of retort found in commercial fish canneries today, is the static batch system for processing cans in saturated steam. A description of the fittings for these retorts is given in the following section; however, many of the other retorting systems referred to above are similar with respect to fittings and methods of operation. The most significant difference between static retorts and continuous systems, is that the latter must have container transfer mechanisms to regulate the movement of cans at a predetermined rate through the heating and cooling sections.

Standard Batch Retorts for Processing Cans in Steam

Irrespective of whether retorts for processing cans in steam are vertical, horizontal or crateless, they have a number of features in common. The major fittings are as follows:

- *Steam inlet:* The steam enters through a perforated steam spreader pipe which provides even distribution of the heating medium throughout the retort. The steam inlet is positioned opposite the main vent: in standard vertical retorts the steam spreader is usually located at the base of the vessel, while in crateless retorts it is circular and at the top; in horizontal retorts it extends the full length of the retort. In general the total cross-sectional area of the perforations in the spreader should be 1.5 to 2 times the smallest cross-sectional area of the steam inlet line.

The steam supply should be capable of bringing a fully loaded retort to operating temperature within 15 min from "steam on" and to regulate temperature to within 1 °C during the process.

- *Vents:* The vent is included to allow the operator to purge all air from the vessel prior to bringing the retort up to operating temperature. It is important that the outlet to the vent is visible so that the operator can see when venting is taking place.

- *Cooling water inlet:* In many retorts the cooling water is supplied via a separate spreader which is positioned at the top of the retort; however, in some installations the cooling water is introduced through the steam spreader at the bottom of the retort but this has the disadvantage that cans at the bottom are cooled first, which causes uneven cooling, particularly in plants where the water pressure is low.

  It is important that the water supply valve can be completely shut off during processing otherwise cold water may leak into the retort and possibly cause under-processing of some cans.

The water supply should be sufficient to fill a fully loaded retort, against the pressure of the steam, within 10 min.

- *Bleeders:* Small bleeders of at least 3 mm diameter: are fitted to the body of the retort and left open (cracked) during the process so that any air and non-condensable gases that are introduced with the steam can be removed.
- *Steam condensate trap:* A steam trap or a bleeder must be fitted to remove condensate which would otherwise accumulate at the base of the retort during heat processing. This is particularly important with crateless retorts, as the random stacking of the cans means that some cans will lie directly on the base of the retort unless they are fitted with false bottoms so that the bottom cans are clear of the condensate.

  The condensate discharge should be positioned so that it can be seen to be functioning correctly by the operator.
- *Pressure safety valves*: All retorts must be fitted with safety valves so that internal pressure does not exceed the recommended working limits.
- *Compressed air line:* Compressed air is used to operate the automatic control valves and for pressure cooling.

  The supply should be sufficient to enable the retort to be cooled with water (within 10 min) without there being any drop in the pressure below that of the steam during the process.
- *Retort baskets and divider sheets:* So that the steam distribution to and around each container is uniform, the retort baskets and the divider sheets must be perforated.
- *Instrumentation:* Retorts require an indicating thermometer (a mercury in glass, or an alternative of comparable accuracy), a recording thermometer (to provide a permanent record and

confirmation that the temperature and the time of the process were as scheduled), and a pressure gauge.

Crateless Batch Retorts for Processing Cans in Steam

Crateless retorting systems are available in which loading, processing, cooling and unloading are all carried out fully automatically, often with the aid of computer control systems which enable one operator to supervise eight to ten retorts simultaneously. Cans are automatically loaded through the top of the retort and fall into the cushion water below. When the prescribed number of containers have been loaded, the oncoming stream of cans is automatically diverted into the next retort, and the top of the retort is closed. Steam is forced into the retort and as it enters it displaces the cushion water through the bottom drain valve.

Once all the air and water have been vented from the system and the retort reaches operating temperature, the process commences. At the completion of the scheduled retorting time, the steam is turned off, and water and compressed air are pumped into the vessel and initial cooling commences. After partial cooling, cans are released into the cushion water canal, which runs underneath the bank of retorts, and from there they are automatically transferred, on conveyors, into the cooling water canal. Although offering considerable labour savings and flexibility, it is important that care be taken while loading and unloading the cans. In the former case there is a danger that, if the retorts are overloaded, or the cushion water level in the retort is too low, or if there are "floaters" (caused by insufficient removal of air prior to sealing the cans), incoming cans will damage the double seams of the uppermost layer of cans. Similarly, during unloading, seam damage can occur if the cans are permitted to drop out of the retorts in an uncontrolled manner.

The risks to the seam are heightened at this stage if the cans are still hot, because the compound will be soft and the cans under positive internal pressure, so that damage to the double seam area may cause momentary venting of the seal. Because of the potential danger to the hermetic seals during unloading, it is strongly recommended that the water level in the cushion water canal be maintained above the level of the exit door.

If this procedure is adopted, the cans gently float down and out of the retort, which means that their double seams are not exposed to as much physical abuse as when they are dropped directly into the cushion water lying below the level of the exit.

Batch Retorts for Processing Flass Containers in Water

The operating principles for processing glass under water in counterbalanced retorts have been discussed. The similarities in retort fittings for processing glass in water and cans in steam are evident when comparing respectively. The main functional characteristics peculiar to systems for processing glass are that:

- water is introduced and mixed with the steam as it enters through spreaders, at the base of the retort (thereby preventing thermal shock breakage); and
- counterbalancing air is required to transmit sufficient pressure through the water to ensure that there is always a greater pressure in the retort than in the container. (it will be recalled that this modification is to prevent the closures from being forced from the finish of the glass during the thermal process).

Hydrostatic Retorts for Processing cans in Steam

In the diagram of the cross section of the hydrostatic retort shown in Figure 34 can be seen the columns of water in the inlet and outlet legs which balance the pressure in the steam dome and give this style of retorts their name.

As the height of the column controls the steam pressure, it also controls the temperature in the steam dome. Cans are automatically loaded onto the chain which carries them through a preheating zone at the top of the inlet leg and down into the column where they are heated by water which becomes progressively hotter the further into the leg they move.

At the bottom of the inlet leg the cans emerge from the water seal and then travel up into the steam dome (or steam chamber). In some hydrostatic retorts the cans have two passes through the dome (one up and the other down), while in others the cans have multiple passes. A two-pass system is illustrated.

The severity of the process depends upon the residence time that the cans are in the dome (which is controlled by chain speed and chain length). And the temperature of the steam (which is controlled by the height of the water column). At the completion of the process the cans move back through the water seal, on the cooling side of the retort, and up into the cooling leg where they are exposed to progressively colder water.

At the top of the cooling leg the cans pass through an air cooling section and then pass down a final cooling section where they are sprayed with cool water. The cooling water canal shown in the illustration is omitted in some hydrostatic retorts.

Because of the high capital investment, the time taken to adjust the conveyor systems to handle different can sizes, and the time required to bring the retorts to operating temperature, hydrostatic cookers are best suited to long I production runs.

When dual chain systems are used, it is possible to process cans I of different sizes simultaneously, for different times but at the same temperature.

While savings of floor space, gentle can handling, and gradual changes in temperature and pressure, are attractive features of these retorts, the systems are expensive to install and maintain, and the costs of breakdowns can be high.

11

Process Control in Fish Canning Operations

Fatal errors in low-acid canned food manufacture are rare, which, given the volume of production, suggests that traditional process control measures (achieved through staff education and training, inspection of facilities and operations and testing or examinations) are effective. This comes as no surprise; for ultimately it is in the canners' interests to assure that their products are not only safe to eat, but also that they are of the expected quality. At the worst, failure to regulate end product quality will lead to outbreaks of food poisoning and expensive recalls; at best, it will gradually undermine the image of the product, and it will limit the ability of the manufacturer to supply to an agreed specification.

The Need for In-process Control

The collective experience of the international fish canning industry is generally sound; nevertheless, manufacturers can ill afford to overlook the outbreaks of botulism which in 1978 and 1932 led to the death of three consumers of commercially canned Alaskan salmon. In both these cases spoilage was the result of post-process contamination by *C. botulinum* (type E). The first outbreak involved only one can from a production lot of 14 600 units, yet the manufacturer inspected (visually and with a dud detector) some 14 million units. Reportedly, 3 515 cans were screened for botulinal toxin and all were negative. The second outbreak was attributed to a single can from a production lot of some 24 000 cans, and led to the recall of 60 million containers from nine canneries. During this investigation approximately 1 000 cans were tested for the presence of botulinus toxin and none was found. In addition to the logistical difficulties of implementing extensive recalls, these incidences demonstrate the impracticability of relying on large scale product recalls and quality audits as a means of detecting unsafe finished product.

In 1978 the International Commission on Microbiological Specifications for Foods (ICMSF), stated that microbiological sampling methods are inappropriate for assessing the safety of low-acid canned foods; they said "...experience demonstrates that, if present, *C. botulinum* would be expected to occur at such low frequency that no conceivable sampling plan would be adequate as a direct measure of its presence." Theoretically, the probability of a single C *botulinum* spore surviving a "botulinum cook" (Fo = 2.8 min) is estimated at 10^{-12}. The probability of botulism arising through C *botulinum*entering containers, via post-process contamination, has been estimated to be from 10^{-7} to 10^{-10}; while that due to botulism being caused by the containers failing to receive a thermal process has been estimated at between 10^{-6} and 10^{-8}.

Notwithstanding that these figures are estimates and difficult to validate, it is clear that when the probabilities of *C. botulinum* (or its toxin) being present in a can of low-acid canned food are so low, the chances of detecting it by terminal analyses are remote - even with a 100% inspection procedure. Should the method of testing be reliable, which is not necessarily the case (e.g., when the presence of vacuum is taken to be the indicator of safety from non-proteolytic type E C *botulinum*), many of the test procedures that are available are destructive and therefore not feasible. Questions as to the value of terminal analyses were revealed at the twelfth session of the Codex Committee on Processed Meats and Poultry Products when it was reported that "...there was general agreement within ICMSF that indirect implant control and hygienic post-processing handling were better measures...(for protecting public health)...than extensive end product examination." (Codex Alimentarius 1982). It is recognized that traditional end product sampling procedures can indicate a gradual deterioration in performance, but they ought not be relied upon to detect manufacturing defects which may compromise the safety of the product.

The two botulism outbreaks cited, and the prevailing attitudes towards traditional terminal analyses (microbiological and/or physical), underscore the desirability of alternate methods to assure the safety of canned fishery products. Hence the attraction of a process control system that minimizes the chances of manufacturing defects, while providing permanent records which demonstrate that the canned product was prepared according to generally recognized standards of good manufacturing practice. It is against this background that application of the Hazard Analysis Critical Control Point (HACCP) concept for in-process control in the manufacture of canned fishery products warrants the attention of canners and regulatory agencies.

The Hazard Analysis Critical Control Point (HACCP) Concept

There are three elements to the HACCP approach for in-process control:

a. Assessment of the hazards associated with the manufacture of the product. In the case of canned fish preserved by heat alone, the hazards are due to the possible survival of, or recontamination by, *C botulinum* or its spores. The risk of botulism arises because:
   - the environment within the can is suitable for toxin formation, and
   - it is conceivable that under some circumstances the finished product is not likely to be treated (e.g., heated prior to consumption) in a manner which can be relied upon to render harmless any toxin that may be present.

b. Identification of the process critical control points (CCPs) to control the hazards. Critical control points are defined as those stages in production where lack of control could lead to the manufacture of an unsafe product due to the presence of organisms of public health significance. This definition places the emphasis on protection of public health; however, some manufacturers broaden their interpretations to include factors which, if not controlled, could affect the marketability of the product. In line with the more stringent approach to CCPs, application of the HACCP concept in canned fish manufacture has the prevention of botulism as its primary objective whereas the wider interpretation encompasses control of safety factors and quality factors.

c. Implementation of standard procedures to monitor the manufacturing process at CCPs. In order to monitor production effectively, the manufacturer must establish performance guidelines against which production at each of the CCPs can be evaluated. Embodied in these guidelines are the quality criteria which determine the specifications for the process. Also, there must be formal procedures to record the results of all the in-process tests which are used to monitor performance, and there must be provision for in line "corrective-action" and "follow -up" control mechanisms. In-process control records should be retained for a period of not less than three years. This is essential to help management regularly review production, but also permanent records are necessary should questions of product suitability arise and a product recall be initiated. In summary this means that at each CCP there must be a specification, a testing procedure, a permanent recording system and a facility for remedial action.

The distinction between traditional control mechanisms and the HACCP approach to in-process control is clear. The former relies on terminal analyses to assess the adequacy of each operation, while the latter relies on in-process testing to demonstrate that all factors critical to the safety (and marketability) of the product have been adequately controlled. Given that improperly manufactured canned fishery products present a potential health risk, and since the safety of high volume canned food production cannot be assessed solely by terminal analyses, the error prevention techniques of the HACCP concept are both rational and potentially more cost effective.

Identification of Critical Control Points

Central to the implementation of the HACCP concept for in-process control is the identification of the critical control points. Since manufacturing techniques for canned fishery products vary greatly between plants, it follows that there will be different CCPs for different production lines. The CCPs for a canning process can be identified with the aid of a process flow diagram which should be constructed for the entire operation from the receipt of the raw and packaging materials through to transport and storage of the finished product.

Process flow diagrams provide a visual means of summarizing the entire sequence of production, and from this it is a simple step to identify CCPs. This list identifies those points in production where the manufacturer must establish specifications, a monitoring system complete with records, and a follow up system to confirm that any adjustments made achieve the desired effects. The frequency of monitoring and the test procedures to be used must also be specified. As the list may not be appropriate for all canning operations, it is important that manufacturers construct their own process flow diagram so that they can be sure no CCPs have been omitted - or included unnecessarily.

Critical Control Point Specifications

In commercial practice there are only a few specifications which are constant for all fish canners. Instead, manufacturers must select specifications which are relevant to each of their particular operations, while taking care that they fulfill the absolute requirement of product safety. Specifications should be set to reflect the desired sensory qualities of the product, they should be realistic and they should be geared to the ability of the plant to match them while remaining profitable. Although it may be difficult to define standards which are applicable to, and accepted by, an entire industry, it is possible to speak in terms of compliance with generally recognized standards of good manufacturing practice (the GMP guidelines referred to throughout. this text, and in particular in sections 1.5, 2.1.4, and 3.2).

For further information regarding GMP guidelines for cannery operations, readers should consult the following publications:

- *FAO/WHO, 1977:* Codex Alimentarius Commission, Recommended international code of practice for canned fish. Rome, FAO/WHO, Joint FAO/WHO Food Standards Programme, CAC/RCP 10-1976: 42 p. Issued also in French and Arabic
- FAO/WHO, 1983: Codex Alimentarius Commission; Recommended international code of practice for low-acid and acidified low-acid canned foods. Rome, FAO/WHO, Joint FAO/ WHO Food Standards Programme, CAC/RCP 23-1979: 50 p.
- *Canned Foods:* Principles of Thermal Process Control, Acidification and Container Closure Evaluation. 4th Edition. Washington: The Food Processors Institute; 1982.

The major objective in seeking compliance with GMP guidelines is protection of public health; and because of this many of the articles referred to above tend to neglect those factors affecting the sensory and the physical properties of the product. Where this information is sought, in the first instance, reference should be made to the Codex Standards for Fish and Fishery Products (CAC/VOL. V- Ed. 1. 1981).

Incubation Tests

Although they should not be used as the sole criterion of product safety, incubation tests can provide valuable information as to the adequacy of the thermal process and also a means of monitoring (indirectly) the microbiological 1 quality of in-coming raw materials. Should a process be of marginal severity, so that a measurable proportion of the population of spore-forming thermophilic bacteria survive, it may be possible to detect changes in the incidence of spoilage after thermophilic incubation tests on the production samples which have been collected as part of routine quality control. It is difficult to predict the level of spoilage in the trade which correlates with a known incidence of spoilage arising from incubation of test samples; however, it was reported by Stumbo (1973) that a thermophilic spoilage level of 1% after thermophilic incubation was found in commercial practice to give rise to a spoilage rate of 0.001% (i.e., 1 in 100 000 units) in the trade. Should a fish canner be able to collect sufficient data to draw their own conclusions concerning the relation between spoilage induced by thermophilic incubation and trade spoilage, the value of these incubation tests becomes clear; particularly for those manufacturers whose products are expected to be marketed in warm climates.

Under normal circumstances there would be little point in routinely monitoring spoilage arising from mesophilic incubation as all the spores which might lead to growth under these conditions should have been eliminated by thermal processes in which target Fo values are 10 to 15 min. However, if there are incidences of mesophilic spoilage detected by these test measures, it is reasonable to conclude that there has been either a significant lapse in the microbiological quality of the raw materials, or a gross failure in the delivery of the scheduled thermal process. In such circumstances corrective remedial action should follow immediately and, suspect stock should be isolated pending a detailed examination.

Incubation test may be carried out in the laboratory or with bulk samples. With the former, the validity of the results must; be verified before any conclusions as to the suitability of the test sample (and by implication, the suitability of the population from which the samples were drawn). Factors to be considered when selecting testing procedures for laboratory incubation include:

- the purpose of the test and its statistical basis;
- the validity of the selection of incubation temperatures;
- the method of examination of incubated containers (e.g., not all spoilage will cause blown cans);
- the sample size required to draw statistically significant conclusions; and
- the tolerable levels for accepting lots with given levels of defectives.

With bulk incubation the factors to be considered include:

- the method can only provide the incidence of blown cans in the lot, under examination;
- the method can highlight changes in spoilage levels, and prompt management to find the causes of any trends; and
- because of the heating lags in bringing cooled cans to incubation temperatures, it is advantageous to commence incubation as soon as possible after the cans leave the retort, when they will still retain some of the heat from the process.

Gibbing

Gibbing is the process of preparing salt herring (or soused herring), in which the gills and part of the gullet are removed from the fish, eliminating any bitter taste. The liver and pancreas are left in the fish during the salt-curing process because they release enzymes essential for flavor. The fish is then cured in a barrel with one part salt to 20 herring. Today many variations and local preferences exist on this process.

History

The process of gibbing was invented by Willem Beuckelszoon (aka Willem Beuckelsz, William Buckels or William Buckelsson), a 14th century Zealand Fisherman. The invention of this fish preservation technique led to the Dutch becoming a seafaring power. Sometime between 1380 and 1386, William Buckels of Biervliet ("Beer Creek") in Zeeland discovered that "salt fish will keep, and that fish that can be kept can be packed and can be exported". Buckels' invention of gibbing created an export industry for salt herring that was monopolized by the Dutch. They began to build ships and eventually moved from trading in herring to colonizing and the Dutch Empire.

The Emperor Charles V erected a statue to Buckels honouring him as the benefactor of his country, and Queen Mary of Hungary after finding his tomb sat upon it and ate a herring. Herring is still very important to the Dutch who celebrate Vlaggetjesdag (Flag Day) each spring, as a tradition that dates back to the 14th century when fishermen went out to sea in their small boats to capture the annual catch, and to preserve and export their catch abroad.

Dried and Salted Cod

Dried and salted cod, often called salt cod or clipfish, is cod which has been preserved by drying after salting. Cod which has been dried without the addition of salt is called stockfish. With the sharp decline in the world stocks of cod due to overfishing, other white fish are often used instead. Sometimes these other species are labeled as such, and sometimes still misleadingly called "salt cod", so the term has become to some extent ageneric name.

Dried and salted cod has been produced in Canada, Iceland, Faroe Islands, Norway and Portugal for over 500 years. It forms a traditional ingredient of the cuisine of many countries around the Atlantic. Traditionally it was dried outdoors by the wind and sun, but today it is usually dried indoors with the aid of electric heaters.

History

The production of salt cod dates back at least 500 years, to the time of the European discoveries of the Grand Banks off Newfoundland. It formed a vital item of international commerce between the New World and the Old, and formed one leg of the so-called triangular trade. Thus it spread around the Atlantic and became a traditional ingredient not only in Northern European cuisine, but also in Mediterranean, West African, Caribbean, andBrazilian cuisines.

The drying of food is the world's oldest known preservation method, and dried fish has a storage life of several years. Traditionally, salt cod was dried only by the wind and the sun, hanging on wooden scaffolding or lying on clean cliffs or rocks near the seaside. Drying preserves many nutrients and is said to make the codfish tastier. Salting became economically feasible during the 17th century, when cheap salt from southern Europe became available to the maritime nations of northern Europe. The method was cheap and the work could be done by the fisherman or his family. The resulting product was easily transported to market, and salt cod became a staple item in the diet of the populations of Catholic countries on 'meatless' Fridays and during Lent.

Names

Dried cod and the dishes made from it are known by many different names. For example, it is known as, *bakaliáros* (Greek), *bacalhau* (Portuguese), *bacalao* (Spanish), *bakaiïao* (Basque), *bacallà* (Catalan), *morue* (French), *baccalà* (Italian), *ráktoguolli/goikeguolli* (Sami), *klippfisk/clipfish* (Scandinavian), *saltfiskur* (Icelandic), *bakalar* (Croatian), and *Saltfish* (Caribbean).

The word compound *bacal-* and its variants are of unknown origin; explorer John Cabot reported that it was the name used by the inhabitants of Newfoundland. When explorer Jacques Cartier ' discovered' the mouth of the St. Lawrence River in what is now Canada and claimed it forFrance, he noted the presence of a thousand Basque boats fishing for cod.

Process

The fish is beheaded and eviscerated, often on board the boat or ship. (This is feasible with whitefish, whereas it would not be with oily fish.) It is then salted and dried ashore. Traditionally the fish was sun-dried on rocks or wooden frames, but today it is mainly dried indoors by electrical heating. It is sold whole or in portions, with or without bones.

Species of Fish

Prior to the collapse of the Grand Banks (and other) stocks due to overfishing, salt cod was derived exclusively from Atlantic cod. Since then products sold as salt cod may be derived from other whitefish, such as pollock, haddock, blue whiting, ling and tusk.

Quality Grades

In Norway, there used to be five different grades of salt cod. The best grade was called superior extra. Then came (in descending order) superior, imperial, universal and popular. These appellations are no longer extensively used, although some producers still make the superior products. The best klippfisk, the superior extra, is made only from line-caught cod. The fish is always of the *skrei*, the cod that once a year is caught during spawning. The fish is bled while alive, before the head is cut off. It is then cleaned, filleted and salted. Fishers and connoisseurs alike place a high importance in the fact that the fish is line-caught, because if caught in a net, the fish may be dead before caught, which may result in bruising of the fillets. For the same reason it is believed to be important that the klippfisk be bled while still alive. Superior klippfisk is salted fresh, whereas the cheaper grades of klippfisk might be frozen first. Lower grades are salted by injecting a salt-water solution into the fish, while superior grades are salted with dry salt. The superior extra is dried twice, much like Parma ham. Between the two drying sessions, the fish rests and the flavour matures.

Culinary Uses

Before it can be eaten salt cod must be rehydrated and desalinated by soaking in cold water for one to three days, changing the water two to three times a day. In Europe the soaked and boiled fish is prepared for the table in a wide variety of ways ; most commonly with potatoes and onions in a casserole or as croquettes. In Jamaica it forms the basis of the national dish ackee and saltfish.

Kipper

A kipper is a whole herring, a small, oily fish, that has been split from tail to head, gutted, salted or pickled, and cold smoked. In the United Kingdom, in Japan, and in some North American regions they are often eaten for breakfast. In the UK, kippers, along with other preserved fish such as the bloater and buckling, were also once commonly enjoyed as a high tea or suppertreat; most popularly with inland and urban working-class populations before World War II.

Terminology

The English philologist and ethnographer Walter William Skeat derives the word from the Old English *kippian*, to spawn. The origin of the word has various parallels, such as Icelandic *kippa* which means "to pull, snatch" and the German word *kippen* which means "to tilt, to incline". Similarly, the English *kipe* denotes a basket used to catch fish. Another theory traces the word kipper to the *kip*, or small beak, that malesalmon develop during the breeding season. As a verb, "to kipper" means to preserve by rubbing with salt or other spices before drying in the open air or in smoke. It is also used in slang to mean being immersed in a room filled with cigarette or other tobacco smoke.

Origin

The exact origin of kippers is unknown, though fish have been slit, gutted and smoked since time immemorial. According to Mark Kurlansky, "Smoked foods almost always carry with them legends about their having been created by accident—usually the peasant hung the food too close to the fire, and then, imagine his surprise the next morning when ...". For instance Thomas Nashe wrote in 1599 about a fisherman from Lothingland in the Great Yarmouth area who discovered smoking herring by accident. Another story of the accidental invention of kipper is set in 1843, with John Woodger of Seahouses in Northumberland, when fish for processing was left overnight in a room with a smoking stove.

These stories and others are known to be apocryphal because the word "kipper" long predates this. Smoking and salting of fish—in particular of spawning salmon and herring which are caught in large numbers in a short time and can be made suitable for edible storage by this practice; predates 19th century Britain and indeed written history, probably going back as long as humans have been using salt to preserve food.

Preparations

"Cold smoked" fish, that have not been salted for preservation, need to be cooked before being eaten safely (they can be boiled, fried, grilled, jugged or roasted, for instance). "Kipper snacks," are precooked and may be eaten without further preparation. In the United Kingdom, kippers are served for breakfast, tea or dinner. In the United States, where kippers are less commonly eaten than in the UK, they are almost always sold as either canned "kipper snacks" or in jars found in the refrigerated foods section. In Haiti, kipper is eaten with scrambled eggs for breakfast or mixed with pasta or rice.

British Variations

Kippers are extremely popular in the Isle of Man. Thousands are produced annually in the town of Peel, where two kipper houses, Moore's Kipper Yard and Devereau and Son, smoke and export herring. Mallaig, once the busiest herring port in Europe, is famous for its traditionally smoked kippers, as well as Stornoway kippers and Loch Fyne kippers.

Related Terms

The Manx word for kipper is *skeddan jiarg* which literally translates as *red herring*. Compare to Irish *scadán dearg*.

A kipper is also sometimes referred to as a "red herring", although particularly strong curing is required to produce a truly red kipper. This term can be dated to the late Middle Ages as quoted here c1400 Femina (Trin-C B.14.40) 27: "He eteþ no ffyssh But heryng red." Samuel Pepys used it in his diary entry of 28 February 1660 "Up in the morning, and had some red herrings to our breakfast, while my boot-heel was a-mending, by the same token the boy left the hole as big as it was before."

Kipper time is the season in which fishing for salmon is forbidden in Great Britain, originally the period 3 May to 6 January, in the River Thames by an Act of Parliament.

Kipper season refers (particularly among fairground workers, market workers, taxi drivers and the like) to any lean period in trade, particularly the first three or four months of the year; possibly a reference to the above usage, or to the need to live frugally during such a period, by (for instance) living on kippers.

Fesikh

Fesikh is a traditional Egyptian fish dish consisting of fermented salted and dried gray mullet, of the mugil family, a saltwater fish that lives in both the Mediterranean and the Red Seas.

The traditional process of preparing it is to dry the fish in the sun before preserving it in salt. The process of preparing fesikh is quite elaborate, passing from father to son in certain families. The occupation has a special name in Egypt, fasakhani.

Fesikh is eaten during the Sham el-Nessim festival, which is a spring celebration from ancient times in Egypt. Egyptians in the West have used whitefish as an alternative. Each year food poisoning tales involving incorrectly prepared fesikh appear in Egyptian periodicals.

Bagoong

Bagoong (Tagalog pronunciation: [bPaoÈoK]) is a Philippine condiment made of partially or completely fermented fish or shrimps and salt. The fermentation process also results in fish sauce (known as *patis*).The preparation of *bagoong* can vary regionally in the Philippines.

Types

Bagoong (spelled as *bugguong* in Ilocano) are usually made from a variety of fish species. Common fishes used include the following:

- Anchovies - locally known as *dilis*, *monamon*, *bolinaw*, or *gurayan* (*Stolephrus* and *Encrasicholina* spp.)
- Round scads - locally known as *galunggong* or *tamodios* (*Decapterus* spp.)
- Bonnetmouths (Redbait or Rubyfish) - locally known as *terong* (*Emmelichthys nitidus*, *Emmelichthys struhsakeri*, and *Plagiogeneion rubiginosum*)
- Ponyfishes - locally known as *sapsap* (*Leiognathus*, *Photopectoralis*, and *Equulites* spp.)
- Rabbitfishes - locally known as *padas* (*Siganus* spp.)
- Bar-eyed gobies - locally known as *ipon* (*Glossogobius giuris*).
- Herrings - *Clupeoides lila*
- Silver perch - locally known as *ayungin* (*Leiopotherapon plumbeus*)

Bagoong made from fish is encompassed by the term *bagoong isda* (literally 'fish *bagoong*') in Luzon and Northern Visayas. In the

Southern Visayas and Mindanao, fish *bagoong* is known as *guinamos* (also spelled *ginamos*). They can be distinguished further by the type of fish they are made of. Those made from anchovies are generally known as *bagoong monamon* or *bagoong dilis* and those from bonnetmouths as*bagoong terong.*

Bagoong can also be made from shrimp fry. This type of *bagoong* is known as *bagoong alamang* or *bagoong aramang* in Ilocano or simply*alamang* or *uyap* in the South.

In rarer instances, it can also be made from oysters, clams, and fish and shrimp roe.A kind of *bagoong* made in the town of Balayan, Batangas is also known as *bagoong Balayan.*

Preparation

Bagoong Isda *and* Bagoong Alamang

Fish *bagoong* is prepared by mixing salt and fish usually by volume; mixture proportions are proprietary depending on the manufacturer. The salt and fish are mixed uniformly, usually by hand. The mixture is kept inside large earthen fermentation jars (known as a *burnay* in Ilocano). It is covered, to keep flies away, and left to ferment for 30-90 days with occasional stirring to make sure the salt is spread evenly. The mixture can significantly expand during the process. The preparation of shrimp *bagoong* is similar, with shrimp cleaned thoroughly and washed in weak brine solution (10%). As in fish *bagoong*, the shrimp are then mixed with salt in a 25% salt to 75% shrimp ratio by weight.

The products of the fermentation process are usually pale gray to white in colour. To obtain the characteristic red or pink colour of some *bagoong*, a kind of food colouring known as *angkak*is added. *Angkak* is made from rice inoculated with a species of red mold (*Monascus purpureus*). High quality salt with little mineral impurities are preferred. High metallic content in the salt used can often result in darker colours to the resulting *bagoong* and a less agreeable undertaste. Likewise, oversalting and undersalting also has a significant impact on the rate and quality of fermentation due to their effects on the bacteria involved in the process. Some manufacturers grind the fermented product finely and sell the resulting mixture as fish paste.

Patis

Patis or fish sauce is a byproduct of the fermentation process, a clear yellowish liquid that floats above the fermented mixture. Sauces similar to *patis* exist throughout Asia and are also used in their respective local cuisines such as *nuoc mam* in Vietnam, *nam pha* in Laos, *hom ha* in China, *nam pla* in Thailand, *shitsuru* in Japan and *saeu chot* in Korea. It has a sharp salty or cheese-like flavor.

To obtain *patis*, fermentation is longer, usually taking 6 months to a year. During longer fermentation processes, the fish or shrimp constituents disintegrate further, producing a clear yellowish liquid on top of the mixture due to hydrolysis. This is the *patis*, it can be harvested once it has developed its characteristic smell.

It is drained, pasteurized, and bottled separately, while the residue is turned into *bagoong*. If the residual solids are not moist enough, brine is usually added. The rate of fermentation can vary depending on the pH levelsof the mixture and the temperature. Exposure to sunlight can also reduce the amount of time required to two months.

Hákarl

Hákarl or kæstur hákarl (Icelandic pronunciation: (Icelandic for "shark") is a food from Iceland. It is a Greenland- or basking shark which has been cured with a particular fermentation process and hung to dry for four to five months. Hákarl is often referred to as anacquired taste and has a very particular ammonia-rich smell and fishy taste, similar to very strong cheese slathered in ammonia.

Hákarl is served as part of a *þorramatur*, a selection of traditional Icelandic food served at *þorrablót* in midwinter. Hákarl is, however, readily available in Icelandic stores all year round and is eaten in all seasons.

Consumption

The Greenland shark itself is poisonous when fresh due to a high content of urea and trimethylamine oxide, but may be consumed after being processed. It has a particular ammonia smell, similar to many cleaning products. It is often served in cubes ontoothpicks. Those new to it will usually gag involuntarily on the first attempt to eat it due to the high ammonia content. First-timers are sometimes advised to pinch their nose while taking the first bite as the smell is much stronger than the taste. It is often eaten with a shot of the local spirit, a type of akvavit, called brennivín. Eating hákarl is often associated with hardiness and strength. It comes in two varieties; chewy and reddish *glerhákarl* (lit. "glassy shark") from the belly, and white and soft *skyrhákarl* (lit. "skyr shark") from the body.

Preparation

Hákarl is traditionally prepared by gutting and beheading a Greenland or basking shark and placing it in a shallow hole dug in gravelly-sand, with the now-cleaned cavity resting on a slight hill. The shark is then covered with sand and gravel, and stones are then placed on top of the sand in order to press the shark. The fluids from the shark are in this way pressed out of the body. The shark ferments in this fashion for 6–12 weeks depending on the season.

Following this curing period, the shark is then cut into strips and hung to dry for several months. During this drying period a brown crust will develop, which is removed prior to cutting the shark into small pieces and serving. The modern method is just to press the shark's meat in a large drained plastic container.

Reactions

Chef Anthony Bourdain, who has travelled extensively throughout the world sampling local cuisine for his Travel Channel show *No Reservations*, has described hákarl as "the single worst, most disgusting and terrible tasting thing" he has ever eaten.

Chef Gordon Ramsay challenged journalist James May to sample three "delicacies" (Laotian snake whiskey, bull penis, and hákarl) on *The F Word*; Ramsay then vomited after eating hákarl, although May kept his down. May's only reaction was, "You disappoint me, Ramsay."

On season 2's Iceland episode of Travel Channel's *Bizarre Foods with Andrew Zimmern*, Andrew Zimmern described the smell as reminding him of "some of the most horrific things I've ever breathed in my life," but said the taste was not nearly as bad as the smell. Nonetheless, he did note that hákarl was "hardcore food" and "not for beginners."

Jeotgal

Jeotgal or *jeot* (Korean pronunciation:) is a salted fermented food in Korean cuisine. It is made with various seafood, such asshrimp, oysters, shellfish, fish, fish eggs, and fish intestines.

Jeotgal is mainly used as a condiment in pickling kimchi and as a dipping sauce for pig's feet (*jokbal*) and blood/noodle sausage (*sundae*). Sometimes *jeotgal*, commonly *saeujeot*, is added to Korean style stews (*jjigae*) and soups (*guk* and *tang*), for flavor instead of using salt orsoy sauce (ganjang).

The types of *jeotgal* vary depending on main ingredients, regions, and family and personal preferences. In past times, due to the limited transportation, regions near seas had more types of *jeot* compared to the inland areas

History

The *Erya* a Chinese dictionary written in the 3rd-5th centuries BC, contains a record about *ji* the origin of *jeotgal*. *Ji* indicates*jeotgal*, food made with fish and is the oldest documents mentioning this food in the historical records.

Types

- Saeujeot - jeot made with small shrimp
- Ojeot saeujeot made with shrimp harvested in May
- Yukjeot saeujeot made with shrimp harvested in June
- Tohajeot jeot made with small fresh water shrimp, which is rare local specialty of Jeolla Province
- Hwangsaegijeot jeot made with fish
- Myeolchijeot - jeot made with anchovies
- Jogijeot - jeot made with croaker
- Jogaejeot jeot made with shellfish
- Guljeot jeot made with oyster
- Eoriguljeot jeot made with salted oysters and hot peppers
- Myeongran jeot jeot made with roe of pollock
- Changnanjeot jeot made with pollock intestines
- Ojingeojeot Jeot made with squid
- Ggolddugijeot jeot made with small squid
- Gejang - jeot made with crabs

Rakfisk

Rakfisk (Norwegian pronunciation: Norwegian fish dish made from trout or sometimes char, salted and fermented for two to three months, or even up to a year, then eaten without cooking.

Origin

The first record of the term *rakfisk* dates back to 1348, but the history of this food is probably even older. No sources are available as to the exact invention year of the rakfisk dish or the fermentation process that produces the raw material for it.

The rakfisk dish is related to the Swedish dish surströmming and probably shares its origin in the ancient Scandinavian culture after the hunting-gathering way of life evolved into a more sophisticated society in which people were able to store food over a considerable period of time.

Etymology

Fisk is the Norwegian word for "fish." *Rak* derives from the word *rakr* in Norse language, meaning "moist" or "soaked" . The word descends from Proto-Indo-European **req,* which means "source" or "drop," and is the root of "rain" and "irrigation."

Preparation Method

Rakfisk is made from fresh trout or char, preferably over 750g. Remove the gills and guts and rinse well so that all the blood is gone. Scrub the blood stripe with a fish brush. Rinse the fish and put it in vinegar solution for about half an hour. Let the fish rest and the vinegar run off for a while. Then place the fish in a bucket, close side-by-side with the abdomen facing up. Fill the abdomen with sea salt (60g per kilogram of fish). Sprinkle tiny amounts of sugar on the fish to speed up the "raking", but not more than a pinch for each layer of fish. Then place the fish under pressure with a lid that fits down into the bucket and a weight on top. The rakfisk bucket is put in a cold place (a stable temperature at about 4 degrees Celsius is the best, but it should be below 8 degrees Celsius at least). After a couple of days you should check if the fish is brined. If not enough fluid has formed to completely cover the fish, add salt brine containing 40g salt per litre of water. The fish may be placed at a higher temperature for some days to make it brine better, but one should be very careful with this.

Leave the rakfisk for two to three months. Rakfisk is well conserved in the brine. When the fish is appropriately "rak," you can put it into a fresh 4% salt brine, which will slow down the "raking" process. Another method for slowing it down is to put the tub in the freezer (or outside if cold enough) for some time. As long as the fish is lying in the brine it will not freeze. Note that all recipes for rakfisk states that the fish must never be in contact with soil. This is very important because of the risk of the wrong bacteria growing in the fish, especially Clostridium botulinum which causes botulism.

Eating

The finished product does not need cooking but is eaten as it is. Rakfisk is traditionally served sliced or as a fillet on flatbrød or lefse and almond potatoes. Some also use raw onion,sour cream, mustard-sauce, a mild form of mustard with dill. Although not an everyday meal, approximately 400 tonnes of rakfisk are produced in Norway yearly.

It is not recommended that rakfisk be eaten by people with a reduced immune defence or by pregnant women.

Shrimp Paste

Shrimp paste or shrimp sauce, is a common ingredient used in Southeast Asian and Southern Chinese cuisine. It is known as terasi (also spelled *trassi*, *terasie*) in Indonesian, ngapi in Burmese, kapi in Thai, Khmer and Lao language, belacan (also spelled*belachan, blachang*) in Malay, mam runc, mam tép and mam tôm in Vietnamese (the name depends on the shrimp used), bagoong alamang (also known as *bagoong aramang*) in Filipino, haam ha/ha jeung in Cantonese Chinese and hom ha/hae ko in Min Nan Chinese.

It is made from fermented ground shrimp, sun dried and either cut into fist-sized rectangular blocks or sold in bulk. It is an essential ingredient in many curries and sauces. Shrimp paste can be found in most meals in Myanmar, Laos, Thailand, Malaysia, Singapore, Indonesia, Vietnamand the Philippines. It is often an ingredient in dipping sauce for fish or vegetables.

Varieties

Shrimp pastes vary in appearance from pale liquid sauces to solid chocolate-coloured blocks. Shrimp paste produced in Hong Kong andVietnam is typically a light pinkish gray while the type used for Burmese, Lao, Cambodian and Thai cooking is darker brown. While all shrimp paste has a pungent aroma, that of higher grades is generally milder. Markets near villages producing shrimp paste are the best places to obtain the highest quality product. Shrimp paste varies between different Asian cultures and can vary in smell, texture and saltiness

Belacan or Terasi

Belacan, a Malay variety of shrimp paste, is prepared from krill, also known as *geragau* in Kristang (Portuguese creole spoken in Malaysia) or*rebon* in Sundanese and Javanese. In Malaysia, normally the krill would be steamed first and after that are mashed into a paste and stored for several months. The fermented shrimp are then prepared, fried and hard-pressed into cakes.

Belacan is used as an ingredient in many dishes, or eaten on its own with rice. A common preparation is *sambal belacan*, made by mixing toasted belacan with chilli peppers, minced garlic, shallot paste and sugar and then fried.

Terasi (trassi in Dutch), an Indonesian variant of dried shrimp paste, is usually purchased in dark blocks, but is also sometimes sold ground. The colour and aroma of terasi varies depending on which village produced it. The colour ranges from soft purple-reddish hue to darkish brown. In Cirebon, a coastal city in West Java, terasi is made from tiny shrimp (krill) called "rebon", the very origin of the city's name. In Sidoarjo, East Java, terasi is made from the mixture of ingredients such as fish, small shrimp (*udang*), and vegetables. Terasi is an important ingredient in Sambal Terasi, also many other Indonesian cuisine, such as sayur asam (fresh sour vegetable soup), lotek (also called gado-gado, Indonesian style salad in peanut sauce), karedok (similar to lotek, but the vegetables are served raw), and rujak (Indonesian style hot and spicy fruit salad).

On the island of Lombok, Indonesia, a more savory and sweet shrimp paste called *lengkare* is made.

Bagoong Alamang

Bagoong alamang is Filipino for shrimp paste, made from minute shrimp or krill (*alamang*) and is commonly eaten as a topping on greenmangoes or used as a major cooking ingredient. Bagoong paste varies in appearance, flavor, and spiciness depending on the type. Pink and salty bagoong alamang is marketed as "fresh", and is essentially the shrimp-salt mixture left to marinate for a few days. This bagoong is rarely used in this form, save as a topping for unripe mangoes. The paste is customarily sauteed with various condiments, and its flavour can range from salty to spicy-sweet. The colour of the sauce will also vary with the cooking time and the ingredients used in the sauteeing. Cincalok is the Malaysian version of 'fresh' bagoong alamang.

Unlike in other parts of Southeast Asia, where the shrimp is fermented beyond recognition or ground to a smooth consistency, the shrimp in bagoong alamang is readily identifiable, and the sauce itself has a chunky consistency. A small amount of cooked or sauteed bagoong is served on the side of a popular dish called "kare-kare", an oxtail stew made with peanuts. It is also used as the key flavouring ingredient of a sauteed pork dish, known as *binagoongan* (lit. "that to which bagoong is applied"). The word *bagoong*, however, is also connoted with the bonnet mouth and anchovy fish version, bagoong terong.

Kapi

In Thailand shrimp paste *(kapi)* is an essential ingredient in many types of *nam phrik*, spicy dips or sauces, and in many Thai curry pastes, such as the paste used in *kaeng som*. Very popular in Thailand is *nam phrik kapi*, a spicy condiment made with fresh shrimp paste and most often eaten together with fried *pla thu* (Short mackerel) and fried, steamed or raw vegetables. In Southern Thailand there are three types of shrimp paste: one made only from shrimp, one containing a mixture of shrimp and fish ingredients, and another paste that is sweet.

Ngapi Yay

A watery dip or condiment that is very popular in Myanmar, especially the Burmese and Karen ethnic groups. The ngapi (either fish or shrimp, but mostly whole fish ngapi is used) is boiled with onions, tomato, garlic, pepper and other spices. The result is a greenish-grey broth-like sauce, which makes its way to every Burmese dining table. Fresh, raw or blanched vegetables and fruits (such as mint, cabbage, tomatoes, green mangoes, green apples, olives, chilli, onions and garlic) are dipped into the ngapi yay and eaten. Sometimes, in less affluent families, ngapi yay forms the main dish, and also the main source of protein.

Hom ha

This Chinese shrimp paste is popular in southeastern China. This shrimp paste is lighter in colour than many southeast Asian varieties and is often used in pork, seafood and vegetable stir fry dishes. The shrimp paste industry has historically been important in the Hong Kong region.

Hae ko or Petis Udang

Hae ko means prawn paste in the Hokkien dialect. It is also called *petis udang* in Malay and Indonesian. This version of shrimp/prawn paste is used in Malaysia, Singapore and Indonesia. In Indonesia it is particularly popular in East Java. This thick black paste has a molasses like consistency instead of the hard brick like appearance of belacan. It also tastes sweeter because of the added sugar. It is used to flavour common local street foods like *popiah* spring rolls, *laksa* curry, *chee cheong fan* rice rolls and *rojak* salads, such as *rujak cingur* and *rujak petis*.

Galmbo

Galmbo is a dried shrimp paste used in Goa, India, particularly in the spicy sauce balchao.

Industry

Shrimp paste continues to be made by fishing families in coastal villages. They sell it to vendors, middlemen or distributors who package it for resale to consumers. Shrimp paste is often known for the region it comes from since production techniques and quality vary from village to village. Some coastal regions in Indonesia such as Bagan Siapi-api in North Sumatra, Indramayu and Cirebon in West Java, and Sidoarjo inEast Java; as well as villages such as Pulau Betong in Malaysia or Ma Wan island in Hong Kong and in Lingayen Gulf, Pangasinan in thePhilippines are well known for producing very fine quality shrimp paste.

Preparation

Preparation techniques can vary greatly; however, the following procedure is most common in China, and much of Southeast Asia.

After being caught, small shrimp are unloaded, rinsed and drained before being dried. Drying can be done on plastic mats on the ground in the sun, on metal beds on low stilts, or using other methods. After several days, the shrimp-salt mixture will darken and turn into a thick pulp. If the shrimp used to produce the paste were small, it is ready to be served as soon as the individual shrimp have broken-down beyond recognition. If the shrimp are larger, fermentation will take longer and the pulp will be ground to provide a smoother consistency. The fermentation/grinding process is usually repeated several times until the paste fully matures. The paste is then dried and cut into bricks by the villagers to be sold. Dried shrimp paste does not require refrigeration.

Availability

Shrimp paste can be found in nations outside Southeast Asia in markets catering to Asian customers. In the Netherlands, Indonesian type of shrimp paste can be found in supermarkets selling Asian foodstuff such as Trassie oedang from the Conimex brand. In the United Statesbrands of Thai shrimp paste such as Pantainorasingh and Tra Chang can be found. Shrimp pastes from other countries are also available in Asian supermarkets and through mail order. It is also readily available in Suriname due to the high concentration of Javanese inhabitants. In Australia shrimp paste can be found in most suburbs where South East Asian people reside.

Surstromming

Surstromming Swedish "soured (Baltic) herring") is a northern Swedish dish consisting of fermented Balticherring. *Surströmming* is sold in cans, which may bulge after prolonged storage, due to the continued fermentation. When opened, the contents release a strong and sometimes overwhelming odor, which explains why the dish is often eaten outdoors. A Japanese study has shown that the smell of a newly opened can of surstromming is the most putrid smell of food in the world, beating similar fermented fish dishes such as the Korean Hongeohoe or Japanese Kusaya.

Origin

A common myth tells that the dish originated with Swedish sailors in the 16th century. The story goes that they only had half the amount of salt needed to keep their fish fresh, so it began to rot. The sailors came across some Finnish islanders and decided to con them by selling the rotten fish to them. The Finns bought it and the sailors went away. A year later the Swedish sailors returned to the island and the locals asked if they had more rotten fish. The sailors decided to try it themselves, liked it and made more.

Another, probably more historically accurate, explanation of the origins of this method of preservation is that it began long ago, when brining food was quite expensive due to the cost of salt. When fermentation was used, just enough salt was required to keep the fish from rotting. The salt raises the osmotic pressure of the brine above the zone where bacteria responsible for rotting (decomposition of proteins) can prosper and prevents decomposition of fish proteins into oligopeptides and amino acids. Instead the osmotic conditions enable the Haloanaerobium bacteria to prosper and decompose the fish glycogen into organic acids, giving it the sour (acidic) properties. Fermented fish is an old staple in European cuisines; for example the ancient Greeks and Romans made a famous sauce from fermented fish called *garum*. Historically, other fatty fish like salmon and whitefish have been fermented not unlike surstromming, and the original gravlax resembled surstromming.

Canning

The herring is caught in spring, when it is in prime condition and just about to spawn. The herring is fermented in barrels for one to two months, then tinned where the fermentation continues. Half a year to a year later, gases have built up sufficiently for the once cylindrical tins to bulge into a more rounded shape. These unusual containers of surstromming can be found in supermarkets all over Sweden. However, certain airlines have banned the tins on their flights, considering the pressurized containers to be potentially dangerous. Species of *Haloanaerobium* bacteria are responsible for the in-can ripening. These bacteria produce carbon dioxide and a number of compounds that account for the unique odour: pungent (propionic acid), rotten-egg (hydrogen sulfide), rancid-butter (butyric acid), and vinegary (acetic acid).

Eating Surströmming

Surströmming is often eaten with a kind of bread known as *tunnbröd,* literally "thin bread". This thin, either soft or crispy bread (not to be confused with crisp bread) comes in big square sheets when crisp or as rounds of almost a metre in diameter when soft.

The custom in *Höga Kusten* ("The High Coast"), the area of northern Sweden where this tradition originates from, is to make a sandwich, commonly known as a "surströmmingsklämma", using two pieces of the hard and crispy kind of *tunnbröd* with butter, boiled and sliced or mashed potatoes (often *mandelpotatis* or almond potatoes) and sliced fish in between and nothing more. In the southern part of Sweden, it is customary to use a variety of condiments such as diced onion, *gräddfil* (fat fermented milk/sour cream) or crème fraîche, chives and sometimes even tomato and chopped dill.

The surströmming sandwich is usually served with snaps and light coloured beers like pilsener or lager. Other drinks of choice are svagdricka (lit. "weak drink", a Swedish low alcohol dark malt beverage brewed since the Middle Ages, slightly similar to porter), water or cold milk. However, exactly what to drink or not to drink to surströmming is highly disputed among connoisseurs. Some claim that cold milk is the right and only choice while others refer to svagdricka as the most traditional drink. Surströmming is usually served as the focus of a traditional festivity, a "surströmmingsskiva" (surströmming party). Many people do not care for surströmming, and it is generally considered to be an acquired taste. Conversely, it is a food which is subject to strong passions (as is lutefisk), and occasionally people like the taste on first try.

Museum

On June 4, 2005, the first surströmming museum in the world was opened in Skeppsmalen, 30 km north of Örnsköldsvik, a town at the northern end of the High Coast.

Controversy

In April 2006, several major airlines (such as Air France and British Airways) banned the fish citing that the pressurised cans of fish are potentially explosive. The sale of the fish was subsequently discontinued in Stockholm's international airport. Those who produce the fish have called the airlines' decision "culturally illiterate," claiming that it is a "myth that the tinned fish can explode."

Because surströmming today contains higher levels of dioxins and PCBs than the permitted levels for fish in the EU, Sweden has had exceptions to these rules. The exception was 2002 to 2011, but an application for renewal of the exemption has been raised to the EU. Producers have claimed that if the application is denied, they will only be allowed to use herring below 17 centimeters, which contain lower levels. This in turn will affect the availability of herring.

In the News

In 1981, a German landlord evicted a tenant without notice after the tenant spread surströmming brine in the apartment building's staircase. In the subsequent trial, the court ruled that the termination was justified when the defendant's (i.e., the landlord's) party demonstrated their case by opening a can inside the courtroom. The court concluded that it "had convinced itself that the disgusting smell of the fish brine far exceeded the degree that fellow tenants in the building could be expected to tolerate."

In August 2011, German food critic and author Wolfgang Fassbender wrote in a major Swiss newspaper, Neue Zürcher Zeitung, that "the biggest challenge when eating surströmming is to vomit only after the first bite, as opposed to before."

By-Product Development

As one of the world's greatest producers of wild caught seafood, Alaska's seafood industry is now turning its attention to developing a seafood waste by-products industry. A significant portion of the fish meal production comes from groundfish operators, either at facilities in Dutch Harbor and Kodiak, or from at sea in factory trawlers.

Based on Commercial Operators Annual Report figures, total pounds of fish meal production in Alaska declined from 111.5 million pounds in 2000, to 68.5 million pounds in 2009. Despite processing 57.4 million pounds less in 2009, the value of the fish meal adjusted to 2009 dollars is about the same, $37.2 million (2000) versus $38.8

million (2009). The volume of fish oil more than doubled and the price climbed nearly fourfold between 2000 and 2009 with eleven million pounds of fish oil worth $2 million processed in 2000 compared with 22.7 million pounds worth $7.5 million in 2009 (inflation adjusted to 2009 dollars).

The number of fish oil processors ranged from a high of 19 in 2001to as few as three in 2008, while the number of fish meal processors declined from a high of 62 in 2001 to 12 in years 2002-2009.

Fish Meal

Introduction

This note briefly describes the manufacture, storage, composition and use of fish meal, and also touches on the problem of air pollution from fish meal plant. The use of fish byproducts for feeding animals is not a new idea; a primitive form of fish meal is mentioned in the Travels of Marco Polo at the beginning of the fourteenth century: '... they accustom their cattle, cows, sheep, camels and horses to feed upon dried fish, which being regularly served to them, they eat without any sign of dislike.'

The utilization of herring as an industrial raw material actually started as early as about 800 AD in Norway. A very primitive process of pressing the oil out of herring by means of wooden boards and stones was employed.

What is Fish Meal?

In the UK the term fish meal means a product obtained by drying and grinding or otherwise treating fish or fish waste to which no other matter has been added. The term white fish meal is reserved for a product containing not more than 6 per cent oil and not more than 4 per cent salt, obtained from white fish or white fish waste such as filleting offal.

These are semilegal definitions, and for convenience fish meal can be defined as a solid product obtained by removing most of the water and some or all of the oil from fish or fish waste. Fish meal is generally sold as a powder, and is used mostly in compound foods for poultry, pigs and farmed fish; it is far too valuable to be used as a fertilizer.

What raw Material is Used?

Virtually any fish or shellfish in the sea can be used to make fish meal, although there may be a few rare unexploited species which would produce a poisonous meal. The nutritional value of proteins from vertebrate fish differs little from one species to another; whole shellfish would however give a nutritionally poorer meal because of the low protein content of the shell.

Most of the world's fish meal is made from whole fish; the pelagic species are used most for this purpose. Where a fishery catches solely for the fish meal industry, it is known as an industrial fishery. Countries with major industrial fisheries are Peru,Norway and South Africa.

Some countries like the UK make fish meal from unsold fish and from offal, that is the heads, skeletons and trimmings left over when the edible portions are cut off. Other countries like Denmark and Iceland use both industrial fish and processing waste.

Fish meal made mainly from filleting offal usually has a slightly lower protein content and a higher mineral content than meal made from whole fish, but a high proportion of small whole fish in the raw material can have the same effect.

The following points are important when selecting species for an industrial fishery:

1. The species must be in large concentrations to give a high catching rate; this is essential because the value of industrial fish is less than that of fish for direct human consumption.
2. The fishery should preferably be based on more than one species in order to reduce the effect of fluctuations in supply of any one species.
3. The total abundance of long lived species varies less from year to year, and
4. Species with a high fat content are more profitable, because the fat in fish is held at the expense of water and not at the expense of protein.

Preservation of the Raw Material

All fisheries experience periods of glut and scarcity, leaving the fish meal factory at times with no raw material to process and at other times with too much. Large amounts of unprocessed material cause storage and odour problems; moreover spoiled material becomes difficult to process and gives a lower yield.

No cheap, completely safe method of preservation has yet been found. Refrigeration is not usually economic, and the known chemical methods of preservation have some disadvantages. Sodium nitrate with formaldehyde is very effective, but unless its addition is very carefully controlled poisonous nitrosamines can be formed when the nitrite reacts with small amounts of trimethylamine in the fish; for this reason nitrite is not used in the UK.

Formaldehyde alone is quite effective in keeping the fish firm enough for processing; it is most useful for species like sand eels that rapidly become semiliquid soon after catching.

Although the addition of about 0-2 per cent by weight of formaldehyde is often enough to provide the required toughening effect, the preservative effect is small at this dilution, and more formaldehyde may make the fish too tough to process.

Bibliography

Allchin, F.R.: *The Agriculture of Early Historic South Asia,* Cambridge University Press, U. K., 1995.

Atre, P. K.: *Fish Genetics and Aquatic Environment,* Navyug, Delhi, 2008.

Bora, K.K. : *Agros Dictionary of Plant Physiology and Biochemistry,* Agrobios, Delhi, 2001.

Breder, C. M., and Rosen, D. E.: *Modes of Reproduction in Fishes, How Fishes Breed,* Natural History Press, New York, 1966.

Carlton, J.T. and R.I. Smith: *Lights' Manual: Intertidal Invertebrates of the Central California Coast,* University of California Press, Berkeley. 1975.

Coultate, T. P. : *Food: The Chemistry of its Components,* London: The Royal Society of Chemistry, 1988.

Deka, Manab: *Fish Fermentation: Traditional to Modern Approaches,* New India Pub, Delhi, 2009.

Devi, Kamla: *Poisonous and Venomous Fishes of Andaman Islands, Bay of Bengal,* Zoological Survey of India, 2003.

Exell A, Burgess PH and Bailey MT: *A-Z of Tropical Fish Diseases and Health Problems,* Howell Book House, New York, 2001.

Farrand, John: *The Audubon Society Master Guide to Birding,* Alfred A. Knopf, New York. 1983

Grizzle J.M.: *Anatomy and Histology of the Channel Catfish,* Auburn Printing Co, 1976.

Hargrove, Mic, David Brown, and Maddy Hargrove: *The Discus: An Owner's Guide to Happy Healthy Fish,* Howell Books, NY, 1999.

Hasler, A. D.: *Orientation and Fish Migration, in Hoar, W. S., and Randall, D. J.,* New York, Academic Press, 1971.

Jee, Chandrawati and Shagufta: *Fish Biotechnology,* A.P.H. Pub, Delhi, 2010.

Kreeger, K.: *Down on the Fish Farm: Developing Effluent Standards for Aquaculture,* Bioscience, Uk, 2000.

LaQue, F.: *Marine Corrosion Causes and Prevention,* John Wiley and Sons, NY, 1975.

Lewbart G.A. *Self-Assesment Color Review of Ornamental Fish,* Iowa State Press,1998.

Mori, Fumitashi: *Aquarium Fish of the World: The Comprehensive Guide to 650 Species*, Chronicle Books, UK, 1991.

Moyle, PB and Cech, JJ: *Fishes, An Introduction to Ichthyology*, Benjamin Cummings, NY, 2003.

Nash, Colin (2011) *The History of Aquaculture* John Wiley and Sons. ISBN 9780813821634.

Pollard, D. A. *Freshwater Fishes of South-eastern Australia.* McDowell, R. M. Reed, Sydney. 1980.

Poppe T.T., *A Color Atlas of Salmonid Diseases,* Academic Press, London, 1996.

Quinn, Thomas P.: *The Behavior and Ecology of Pacific Salmon and Trout,* University of Washington Press, Washington, 2005.

Rebelin W.E.: *The Pathology of Fishes,* The University of Wisconsin Press, UK, 1975.

Reichenbach-Klinke H. H.: *Fish Pathology,* T.F.H. Publications, Inc. Neptune City, NJ. 1973.

Selvamani, B.R. and R.K. Mahadevan: *Fish and Fishery Culture: Assessment and Evaluation,* Campus Books International, Delhi, 2008.

Sharma, L.L. and A.K. Mathur: *Hand Book of Freshwater Ornamental Fishes,* Yash Pub house, Delhi, 2006.

Srivastava, M.P.: *Natural History of Fishes and Systematics of Freshwater Fishes of India,* Narendra Publishing House, Delhi, 2002.

Tietenberg TH: *Environmental and Natural Resource Economics: A Contemporary Approach,* Pearson/Addison Wesley, NY, 2006.

Turton, John: *The Angler's Manual; or Fly-fisher's Oracle.* London: R. Groombridge, 1836.

Untergasser D.: *Handbook of Fish Diseases,* T.F.H. Publications, Neptune, NJ, 1989.

Ward, Barbara E.: *Varieties of the Conscious Model: Fishermen of South China,* Tavistock, London, 1965.

Wright, Leonard M. Jr.: *Fishing the Dry Fly As A Living Insect-An Unorthodox Method.* New York: E. P. Dutton & Co., 1972.

Yohei, Sakamoto and Mori, Fumitashi: *Aquarium Fish of the World: The Comprehensive Guide to 650 Species,* Chronicle Books, UK, 1991.

Index

I

K

L

M

N

O

P

Q

R

S

T

V

W

❑❑❑